Math Kangaroo Competition
Problems and Solutions
Grades 5 and 6
Volume I

Problems and Solutions
Even Years
1998–2020

Editor in Chief
Agata Gazal
Chief Editorial Officer for Math Kangaroo
Bozeman, MT

Reviewers
Joanna Matthiesen
Chief Executive Officer for Math Kangaroo
Granger, IN

Izabela Szpiech
Chief Financial Officer for Math Kangaroo
Norridge, IL

Kasia Nalaskowska
Chief Information Officer for Math Kangaroo
Aurora, IL

Svetlana Savova
Chief Academic Officer for Math Kangaroo
Johns Creek, GA

Contributors
Maria Omelanczuk
Former CEO and President for Math Kangaroo
Oswego, IL

Andrzej Zarach, PhD
Math Content Reviewer, Professor Emeritus of East Stroudsburg University
East Stroudsburg, PA

Dawid Zarach
Math Content Reviewer
East Stroudsburg, PA

Cover and Graphics Credit
Magdalena Teodorowicz
Chief Design Officer for Math Kangaroo
Cordova, TN

Agata Gazal
Chief Editorial Officer for Math Kangaroo
Bozeman, MT

We would like to give special thanks to other countless people who contributed to the questions and solutions of this book since 1998, chiefly to the Math Kangaroo question writers from all over the world that are part of the AKSF organization (www.aksf.org); Math Kangaroo solution writers also include Math Kangaroo USA competition organizers and Math Kangaroo Alumni (www.mathkangaroo.org). We would also like to thank the hundreds of educators who gave us feedback on the questions and solutions, and finally the tens of thousands of students that take the challenge each year. Thank you all for your help in developing this book.

www.mathkangaroo.org

Printed by:
Classic Printing & Thermography
Wood Dale, IL

For additional copies of this book, please contact the publisher:
Math Kangaroo USA
info@mathkangaroo.org

ISBN 978-0-9814868-1-9

Preface

Many people enjoy the challenge of solving math riddles and other types of puzzles. This book presents 360 entertaining problems and solutions presented to 5th and 6th grade students during the Math Kangaroo Competition even years spanning 1998-2020, the total of 12 tests. Each test consists of 30 questions divided into easy, medium, and difficult categories. The questions were selected at the annual *Kangourou sans Frontières* meeting where mathematicians from over 80 countries work together to choose the most engaging and age-appropriate questions for the annual Math Kangaroo competition.

This easy-to-use resource book includes fun questions, pictures, and interesting solutions that challenge children to use math and logic as a tool for understanding the world around them. Problem solving is a skill that all children use, sometimes without even knowing it. This book will help students practice their math skills that often involve logical reasoning and reflecting on the solutions.

We hope this book will be cherished not just by students who love mathematics but also by teachers who are passionate about teaching unconventional and challenging math. We believe students will benefit from this book and find it both insightful and entertaining.

Joanna Matthiesen

President and CEO of Math Kangaroo USA

Table of Contents

Part I: Problems

Problems from Year 1998

Problems 3 points each

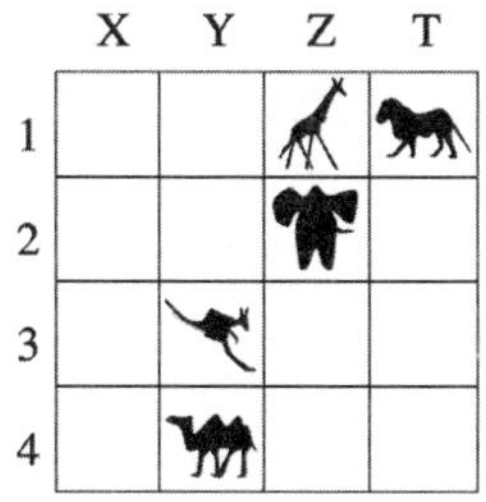

1. Give the coordinates of the kangaroo.

(A) 2X (B) 3Y (C) 1Y (D) 4Z (E) 3T

2. A kangaroo is traveling from START to FINISH using the paths shown in the picture. Each segment is marked with the time (in minutes) which the kangaroo needs to travel that segment. What is the shortest time needed for the kangaroo to reach FINISH?

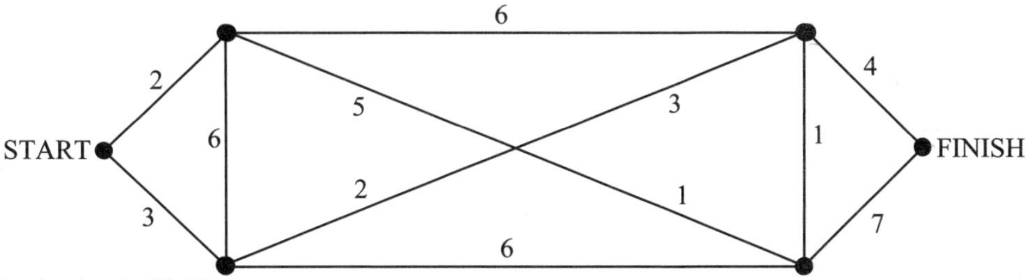

(A) 11 minutes (B) 13 minutes (C) 16 minutes (D) 19 minutes (E) 12 minutes

3. Among the puzzle pieces below, two have the same area. Which two?

(A) 4 and 2 (B) 1 and 5 (C) 1 and 3 (D) 4 and 5 (E) 3 and 5

4. What number is the smallest natural number greater than 360 and at the same time a square of a natural number?

(A) 400 (B) 362 (C) 361 (D) 900 (E) other number

5. One night and day period on Mars is 40 minutes longer than on Earth. How much longer is a week on Mars than on Earth?

(A) 4 h 40 min (B) 2 h 80 min (C) 7 h 20 min (D) 40 min (E) 0 min

6. How many rectangles are there in the picture?

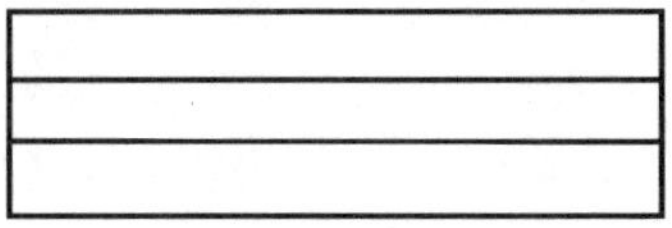

(A) 1 (B) 3 (C) 4 (D) 5 (E) 6

7. A wall clock strikes at every hour (the number of strikes corresponds to the time, so for example, at 10 a.m. and at 10 p.m. you will hear 10 strikes). The clock also strikes once at the half-hour mark. How many strikes can be heard in one 24-hour period?

(A) 24 (B) 136 (C) 180 (D) 196 (E) 240

8. It is now the spring of 1998. The last Summer Olympics took place in 1996, and the last Winter Olympics finished just a few weeks ago. Both the Summer and Winter Olympics take place every 4 years. Counting both the summer and winter competitions, how many more times will the Olympics take place before March 20, 2051?

(A) 13 (B) 16 (C) 25 (D) 26 (E) other answer

9. In how many ways can you place two identical 1 dollar coins in three pockets?

(A) 2 (B) 3 (C) 4 (D) 6 (E) 8

10. Andy is wearing a t-shirt with the word KANGUR on it. He is standing in front of a mirror. What word does he see when he looks at his t-shirt in the mirror?

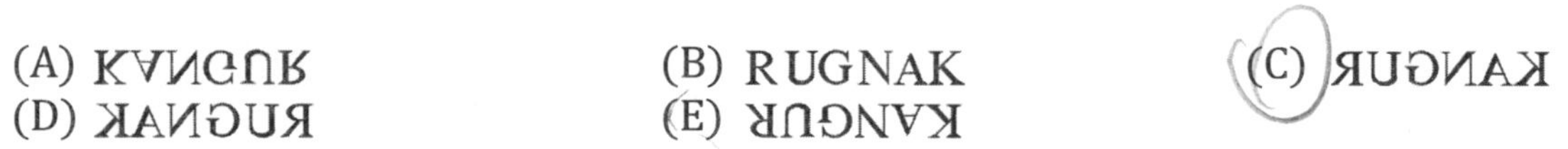

Problems 4 points each

11. What number is at the top of the pyramid if it is formed according to the pattern shown below?

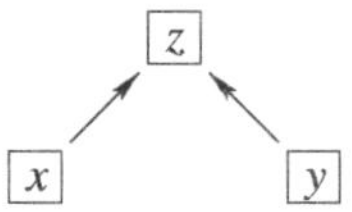

$$z = \frac{x + y}{2}$$

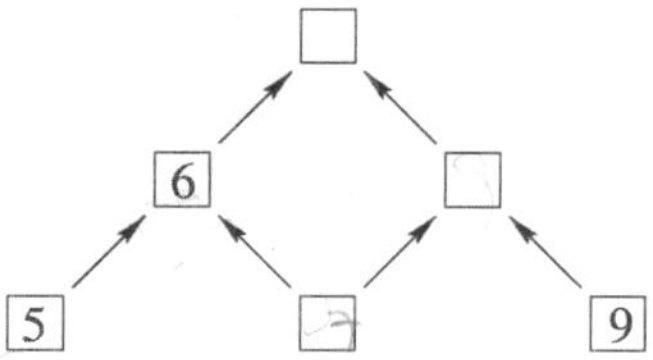

(A) 5 (B) 7 (C) 8 (D) 9 (E) 12

12. A watermelon weighs $\frac{4}{5}$ kg more than $\frac{4}{5}$ of the same watermelon. How much does the watermelon weigh?

(A) $\frac{8}{5}$ kg (B) 4 kg (C) 3 kg (D) 4.5 kg (E) 5 kg

13. There are stools and chairs in the room. Each stool has 3 legs and each chair has 4 legs. Altogether there are 17 legs. How many chairs are there in the room?

(A) 5 (B) 4 (C) 3 (D) 2 (E) 1

14. If $\square + \bigcirc = 30$, $\square + \triangle + \triangle = 160$, and $\triangle + \bigcirc = 80$, then $\square + \triangle + \bigcirc + \bigcirc = ?$

(A) 80 (B) 100 (C) 110 (D) 210 (E) 90

15. When from any three-digit number we subtract that number written backwards, the difference will always be a number that is divisible by:

(A) 7 (B) 2 (C) 5 (D) 9 (E) 13

16. When Mr. Kowalski was asked how old he was, he said, "I have lived 44 years, 44 months, 44 weeks, 44 days, and 44 hours." How many years old is Mr. Kowalski?

(A) 44 (B) 47 (C) 48 (D) 49 (E) 50

17. There are 3 married couples. In how many ways can we form a three-person group in which there will not be a married couple?

(A) 1 (B) 2 (C) 6 (D) 8 (E) 20

18. On Monday morning, a snail fell down a well which is 10 meters deep. During the day, it climbs up 2 meters, and during the night it slides down 1 meter. On what day of the week will the snail get out of the well?

(A) Tuesday (B) Thursday (C) Saturday (D) Sunday (E) Monday

19. John and Stan each have three cards marked with digits. John's cards are marked with the digits 2, 4, and 6, and Stan's cards are marked with the digits 1, 3, and 5. They are taking turns placing their cards in this diagram: $\square\square\square\square\square\square$. John will fill in the first spot on the left, Stan the second spot, and so on. John is trying to make the final number as small as possible, and Stan is trying to make it as large as possible. What number will they form?

(A) 123456 (B) 654321 (C) 254361 (D) 253146 (E) 253416

20. At a fair, the tickets for four various types of rides cost 2 dollars, 3 dollars, 4 dollars, and 5 dollars, respectively. A class took a field trip to the fair. They bought enough tickets for each student to go on each of the four rides once. The tickets cost 280 dollars altogether. How many tickets did they buy?

(A) 14 (B) 20 (C) 40 (D) 80 (E) 140

Problems 5 points each

21. A juice carton which is $\frac{3}{4}$ full contains enough juice to fill $1\frac{1}{2}$ glasses. How many glasses will the juice from 5 full cartons fill?

(A) $7\frac{1}{2}$ (B) $3\frac{3}{4}$ (C) 8 (D) 10 (E) $8\frac{1}{4}$

22. Whole numbers from 1 to 12 are placed in the figure in such a way that the sum of the four numbers found along each segment is the same (see the picture). Under which letter is the number 7 hidden?

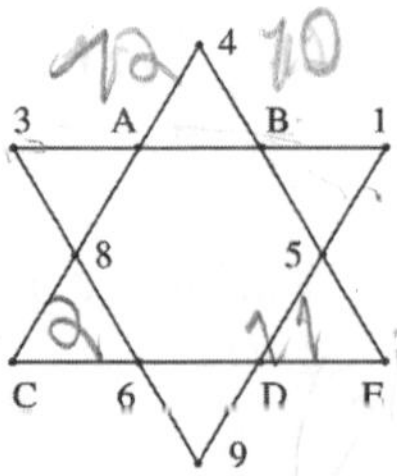

(A) A (B) B (C) C (D) D (E) E

23. What is the ones digit in the number $1^2 + 2^2 + 3^2 + 4^2 + 5^2 + 6^2 + 7^2 + 8^2 + 9^2 + 10^2$?

(A) 1 (B) 3 (C) 5 (D) 7 (E) 9

24. Choose the order in which it would be impossible to place the blocks in the box.

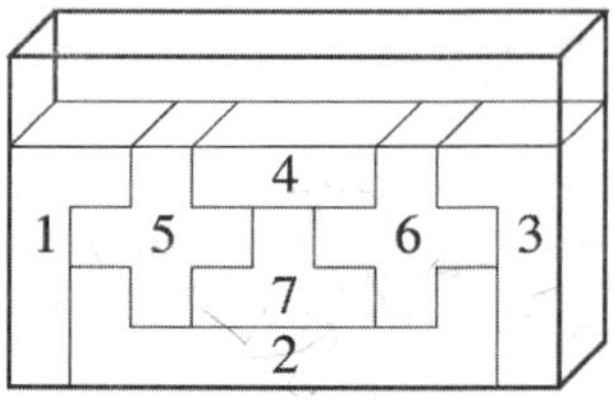

(A) 2, 7, 5, 6, 4, 1, 3 (B) 2, 7, 5, 1, 6, 4, 3 (C) 2, 7, 6, 3, 4, 5, 1
(D) 2, 7, 6, 5, 3, 1, 4 (E) 2, 7, 5, 1, 6, 3, 4

25. In the diagram below, house X is shown 4 times (from different sides), and house Y is shown just once. Which picture shows house Y?

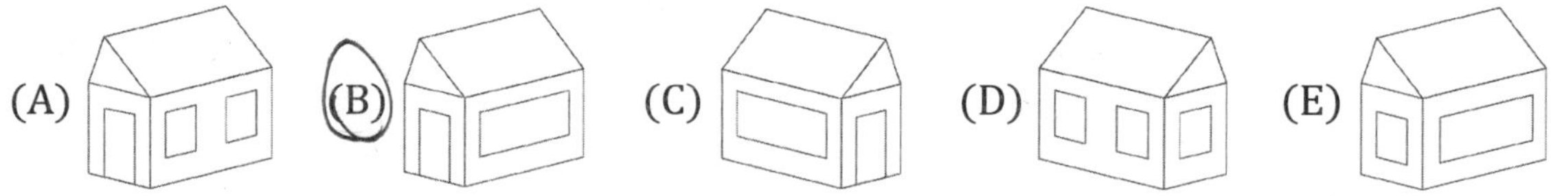

26. Four teams took part in a soccer tournament. The rules were:
a) each team plays against each of the other teams exactly once, and
b) a team gets 3 points for winning, 0 points for losing, and 1 point if there is a tie.
At the end of the tournament, the teams had 5 points, 3 points, 3 points, and 2 points, respectively. How many of the games ended in a tie?

(A) 1 (B) 2 (C) 3 (D) 4 (E) 5

27. Ten coins were placed as shown in the picture. What is the smallest number of coins that need to be removed so that an equilateral triangle cannot be formed by the centers of any three of the remaining coins?

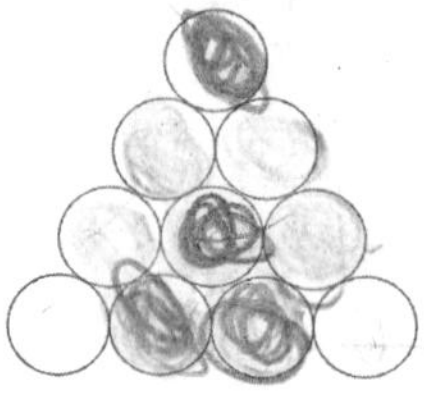

(A) 3 (B) 4 (C) 5 (D) 6 (E) 7

28. Snow White lined up the Seven Dwarfs from shortest to tallest. She divided 77 berries among them. The shortest dwarf got a certain number of berries, the next one got one berry more, and so on. How many berries did the tallest dwarf get?

(A) 17 (B) 8 (C) 14 (D) 10 (E) 15

29. In the basketball semi-finals, team A plays against team B and team C plays against team D. The winners of these two games will play against each other for first and second place, and the losers will play for third and fourth place. How many possible outcomes are there?

(A) 4 (B) 8 (C) 12 (D) 16 (E) 24

30. The area of the shaded triangle is one-fourth of the area of the rectangle (see the picture). What part of the length of the base of the rectangle is the base of the triangle?

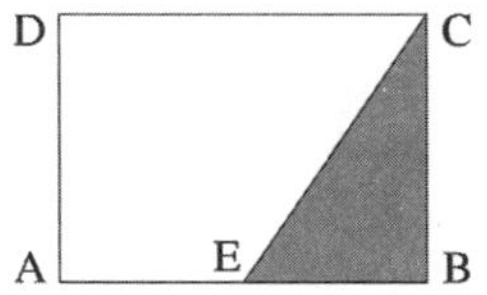

(A) $\frac{1}{2}$ (B) $\frac{1}{3}$ (C) $\frac{1}{4}$ (D) $\frac{3}{4}$ (E) $\frac{2}{3}$

Problems from Year 2000

Problems 3 points each

1. There are 29 students in a class, and the number of girls is 3 greater than the number of boys. How many girls are in this class?

(A) 6 (B) 13 (C) 16 (D) 19 (E) 15

2. The number $-11-2(-7)$ is equal to

(A) 3 (B) -3 (C) -25 (D) 25 (E) 16

3. There are 585 drawers in an old, huge chest. In each drawer there are 3 mice, and each one is the mother of 5 little mouse pups, which also live with them. How many little mouse pups are there in the chest?

(A) $(585+3)+3$ (B) $(585\times 3)+5$ (C) $(585\times 5)+3$
(D) $585\times 3\times 5$ (E) $585\times(3+5)$

4. The sum of five consecutive natural numbers is equal to 2000. The greatest of these numbers is

(A) 490 (B) 475 (C) 471 (D) 423 (E) 402

5. A train is 56 km away from the nearest station and is approaching the station traveling 9 km for every 10 minutes. How far from the station will the train be after 30 minutes?

(A) 47 km (B) 39 km (C) 31 km (D) 29 km (E) 26 km

6. The picture shows the mirror reflection of a wall clock. What time does the actual clock show?

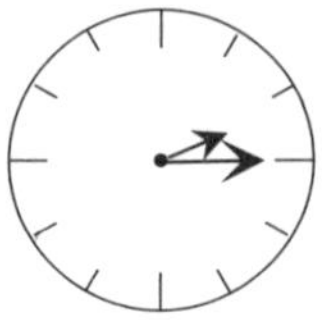

(A) 3:15 (B) 10:15 (C) 10:45 (D) 2:15 (E) 9:45

7. It is now the year 2000. How many twos and fives are there among the prime factors of the number 2000?

(A) 2 twos and 5 fives (B) 3 twos and 3 fives (C) 3 twos and 4 fives
(D) 4 twos and 3 fives (E) 4 twos and 4 fives

8. Anna's birthday present is placed in a box with dimensions 10 cm × 10 cm × 30 cm and wrapped with a ribbon as shown in the picture. What is the length of the ribbon (ignore the length needed to make knots)?

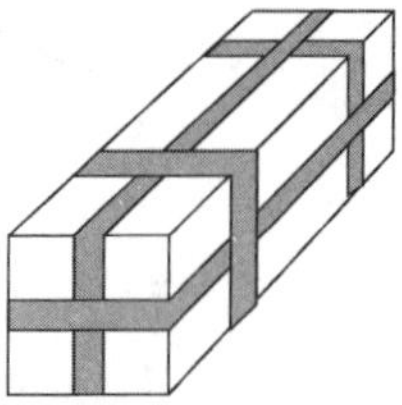

(A) 200 cm (B) 240 cm (C) 260 cm (D) 300 cm (E) 250 cm

9. One of the corners of a square was folded (see the picture). Two other corners of the square were then folded the same way. How many vertices does the figure formed in this way have?

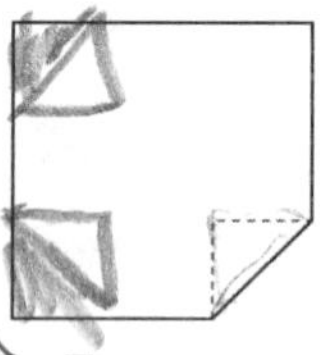

(A) 3 (B) 4 (C) 5 (D) 6 (E) 7

10. The kangaroo is facing towards X (see the picture). Which letter will the kangaroo face if it rotates 270° clockwise?

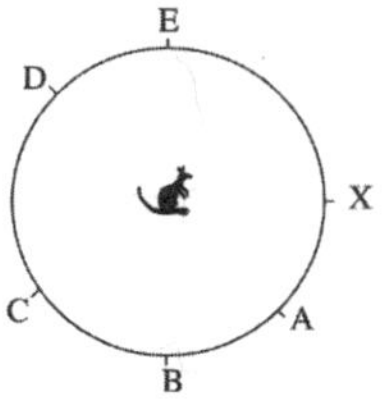

(A) A (B) B (C) C (D) D (E) E

Problems 4 points each

11. How many two-digit numbers are divisible by both 2 and 7?

(A) 8 (B) 7 (C) 6 (D) 5 (E) 4

12. If $A - 1 = B + 2 = C - 3 = D + 4 = E - 5$, then which of the numbers A, B, C, D, and E, is the greatest?

(A) A (B) B (C) C (D) D (E) E

13. Out of how many small squares will a figure be made, if it is made the same way as the figure in the picture but has 10 steps?

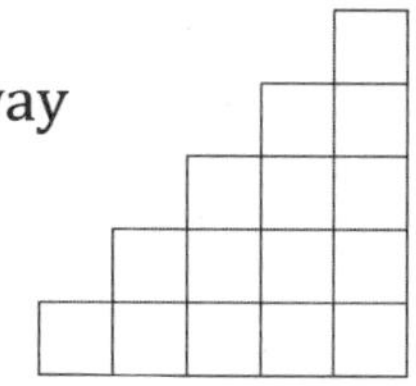

(A) 25 (B) 30 (C) 40 (D) 55 (E) 100

14. How long will it take to print one million forms, if it takes 1 minute to print 100 of these forms?

(A) 160 hours and 40 minutes
(B) 166 hours and 40 minutes
(C) 120 hours and 40 minutes
(D) 18 hours and 10 minutes
(E) 120 hours

15. Each of the five neighbors owns a rectangular plot of land with the same area. The flower gardens of their land are fenced in (solid line in the pictures). Who has the longest fence?

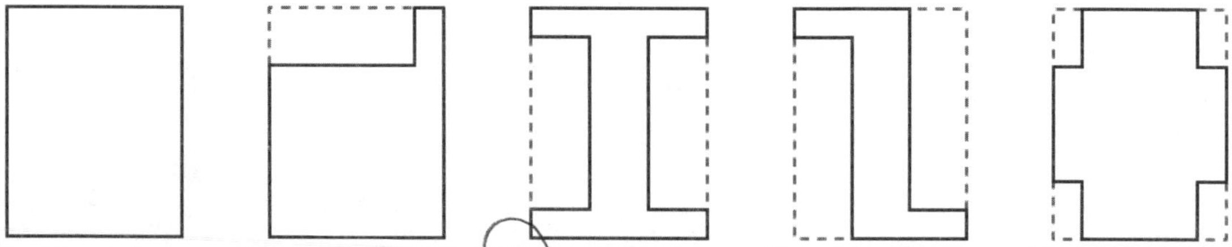

(A) Mr. Adam (B) Mr. John (C) Mr. Jack (D) Mr. Peter (E) Mr. Mark

16. Number a is greater than number b. The difference between numbers a and b is 15. If we decrease number a by 5 and increase number b by 2, then the difference will

(A) decrease by 8
(B) decrease by 5
(C) increase by 8
(D) increase by 5
(E) decrease by 7

17. John comes to the computer lab every day, Karl every 2 days, Stan every 3 days, Adam every 4 days, Paul every 5 days, and Peter every 6 days. Today they are all at the computer lab. In how many days will they all be there together again?

(A) 6 days (B) 20 days (C) 30 days (D) 60 days (E) 90 days

18. What is the sum of the areas of all the triangles you can see in the picture to the right?

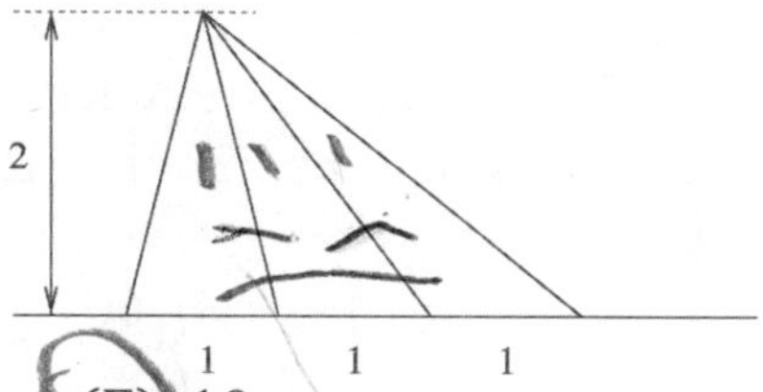

(A) 4 (B) 5 (C) 7 (D) 8 (E) 10

19. Which four digits need to be removed from the number 4921508 to get the smallest possible three-digit number?

(A) 4, 9, 2, 1 (B) 4, 2, 1, 0 (C) 1, 5, 0, 8 (D) 4, 9, 2, 5 (E) 4, 9, 5, 8

20. How many four-digit numbers with a sum of their digits equal to 3 are there?

(A) 6 (B) 8 (C) 9 (D) 10 (E) 12

Problems 5 points each

21. The leaders of the math camp in Zakopane decided to divide 96 participants into equal groups where the number of participants in each group is between 5 and 20. In how many ways can this be done?

(A) 10 (B) 8 (C) 5 (D) 4 (E) 2

22. The scale shown in the picture is in balance. On the scales there is a 20 g weight and certain solids: cubes and cylinders. All the solids (cubes and cylinders) together weigh 500 g. How much does one cube weigh?

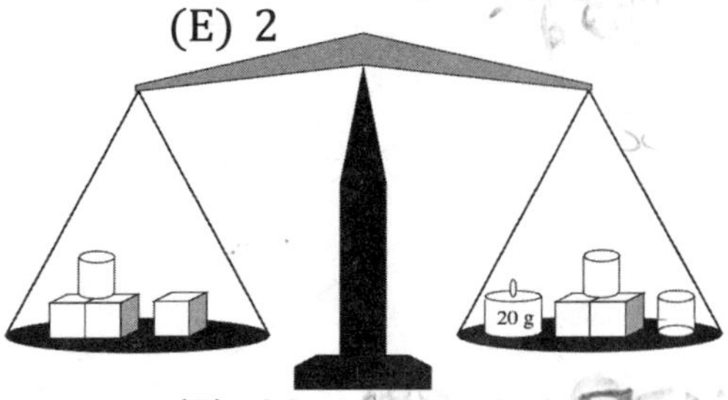

(A) 40 g (B) 50 g (C) 60 g (D) 70 g (E) 80 g

23. The length of one of the sides of a rectangle was increased by 10%, and the length of the other side of the rectangle was decreased by 10%. How did the area of the rectangle change?

(A) It did not change. (B) It decreased by 1%. (C) It increased by 1%.
(D) It increased by 20%. (E) It depends on the lengths of the sides.

24. What is the area of the shaded figure (see the picture to the right)?

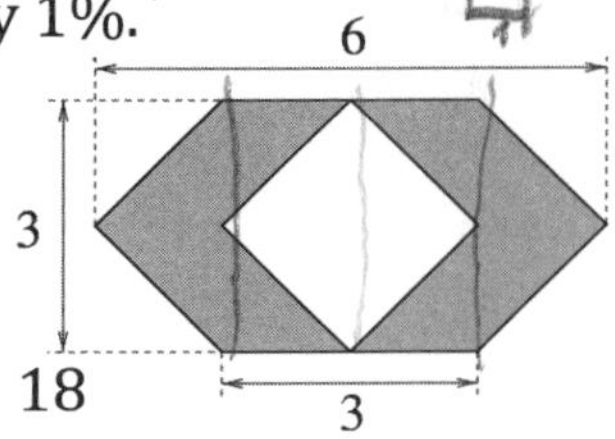

(A) 5 (B) 9 (C) 12 (D) 15 (E) 18

25. The jump of a little kangaroo is 1 meter long and takes one-half of a second. His mother's jump is 3 meters long and takes one second. The mother and the little kangaroo start at the same time from the same place and are jumping towards a eucalyptus which is 180 meters away. For how many seconds will the mother be waiting for the little kangaroo at the eucalyptus?

(A) 30 (B) 60 (C) 10 (D) 120 (E) 20

26. In the pentagon ABCDE (see the picture), the measure of angle BAC is

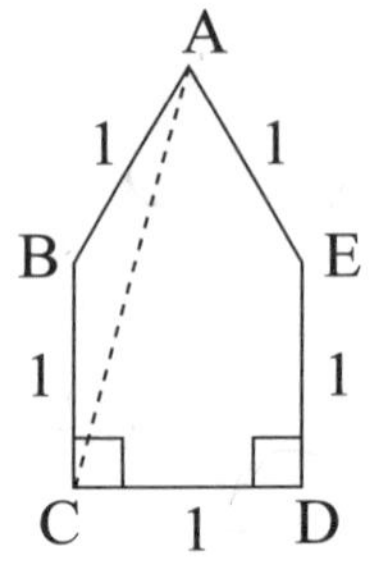

(A) 15° (B) 12° (C) 30° (D) 20° (E) other

27. How many different counterweights can be found to balance a scale using any or all of the following three weights: 1 kg, 3 kg, and 9 kg? (We can place any weight on any side.)

(A) 3 (B) 6 (C) 11 (D) 13 (E) 14

28. We used 8 grams of playdough to make a cube with an edge of 2 cm. How many grams of playdough do we need to make a cube with an edge of 4 cm?

(A) 16 (B) 24 (C) 32 (D) 48 (E) 64

29. The body of a certain caterpillar is made up of five spherical parts, 3 of which are yellow and 2 are green. What is the greatest possible number of different types of this caterpillar that could exist?

(A) 6 (B) 8 (C) 9 (D) 10 (E) 12

30. We have 3 boxes in one row: one red, one green, and one blue. We also have 3 objects: a coin, a shell, and a bead. In each of the boxes there is only one of the objects. We know that:

- the green box is to the left of the blue box;
- the coin is to the left of the bead;
- the red box is to the right of the shell; and
- the bead is to the right of the red box.

In which box is the coin?

(A) in the red box (B) in the green box (C) in the blue box
(D) This cannot be determined.
(E) The conditions given above cannot all be true at the same time.

Problems from Year 2002

Problems 3 points each

1. The number 2002 read from left to right and from right to left is the same. Which number from the numbers below does not have this property?

(A) 1991 (B) 2323 (C) 2112 (D) 2222 (E) 4334

2. The picture below is a sketch of a castle. Which of the lines below does not belong to the sketch?

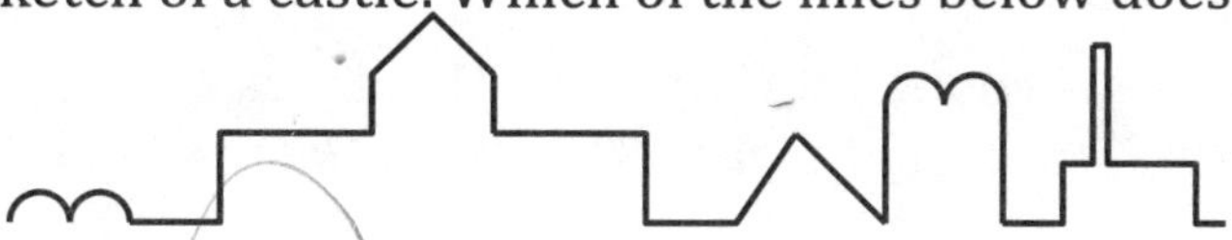

(A) 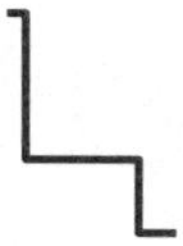 (B)  (C)  (D)  (E) 

3. Mr. and Mrs. Kowalski have three daughters. Each of them has two brothers. How many children does the Kowalski family have?

(A) 9 (B) 7 (C) 6 (D) 5 (E) 11

4. In which number below is the square of tens digit equal to the triple of the sum of the digits of hundreds and ones?

(A) 192 (B) 741 (C) 385 (D) 138 (E) 231

5. The product $2^2 \times 2^{2000} \times 2$ is equal to:

(A) 2^{400} (B) 2^{2002} (C) 2^{2003} (D) 2^{4002} (E) 2^{4001}

6. On which string is the number of black hearts equal to two thirds of the number of all the hearts on the string?

(A)  (B)  (C) 

(D)  (E) 

7. Which of the numbers below is the greatest?

(A) $10 \times 0.001 \times 100$ (B) $0.01 \div 100$ (C) $100 \div 0.01$
(D) $10{,}000 \times 100 \div 10$ (E) $0.1 \times 0.01 \times 10{,}000$

8. 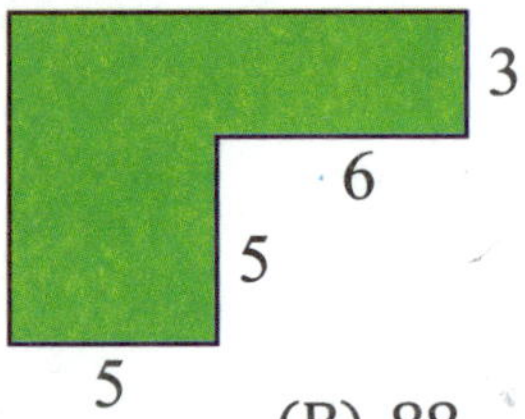

The area of the figure in the picture is equal to:

(A) 43 (B) 88 (C) 58 (D) 30 (E) 15

9. The area of a certain rectangle is equal to 1 m^2. What is the area of a triangle that was cut off from the rectangle along the line connecting the midpoints of the two adjacent sides? (1 m = 10 dm = 100 cm)

(A) 33 dm^2 (B) 25 dm^2 (C) 40 dm^2 (D) 3,750 cm^2 (E) 1,250 cm^2

10. We subtracted the smallest three-digit number with all different digits from the greatest three-digit number with all different digits. The result was:

(A) 864 (B) 885 (C) 800 (D) 899 (E) a different number

Problems 4 points each

11. Figures I, II, III, and IV are squares. The perimeter of square I is equal to 16 m, and the perimeter of square II is equal to 24 m. The perimeter of square IV is equal to:

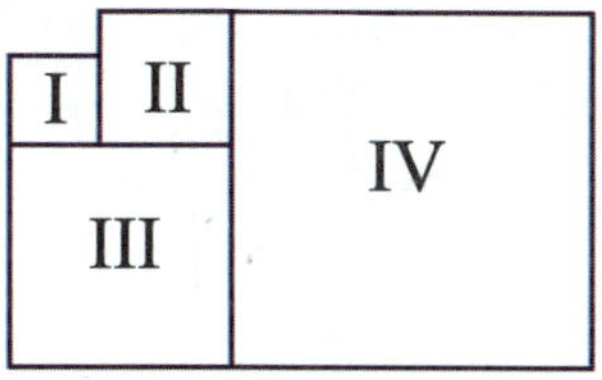

(A) 56 m (B) 60 m (C) 64 m (D) 72 m (E) 80 m

12. One medal can be cut out from a gold square plate. If four medals are made from four plates, the remaining pieces of those four plates can be used to make one more plate. What is the largest number of medals that can be formed when 64 plates are used?

(A) 85 (B) 64 (C) 80 (D) 84 (E) 100

13. Rectangle ABCD (see the picture) is made of 24 little squares with the length of each side equal to 1. What is the area of the triangle ALM?

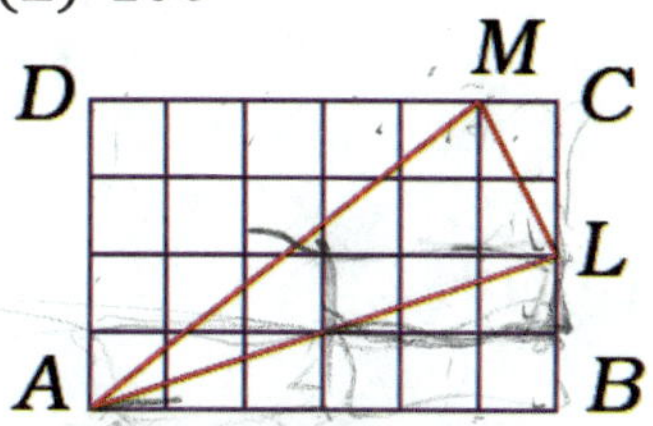

(A) 5 (B) 6 (C) 7 (D) 8 (E) none of these

14. In the picture below the coordinates of points A and B are indicated. What are the coordinates of points C and D if $|AB| = 2|BC|$ and $|BC| = 2|CD|$?

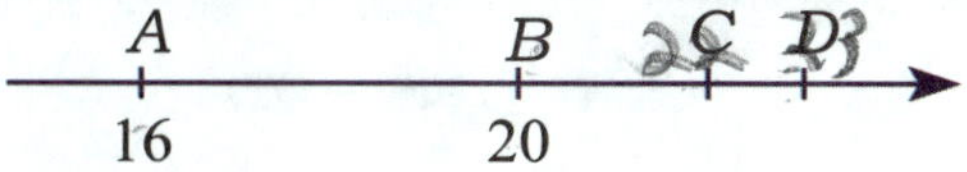

(A) 24 and 32 (B) 24 and 28 (C) 24 and 26 (D) 22 and 24 (E) 22 and 23

15. Mark has 9 sticks with the lengths of 1 dm, 2 dm, 3 dm, 4 dm, 5 dm, 6 dm, 7 dm, 8 dm, and 9 dm. Using the sticks, he builds triangles with each side made with one stick. How many triangles with a side of 1 dm can be built using these sticks?

(A) 6 (B) 3 (C) 2 (D) 1 (E) 0

16. How many convex angles with different measures are made by the rays with P as the starting point (see picture)? (A convex angle is an angle that is larger than a zero angle and smaller than a straight angle.)

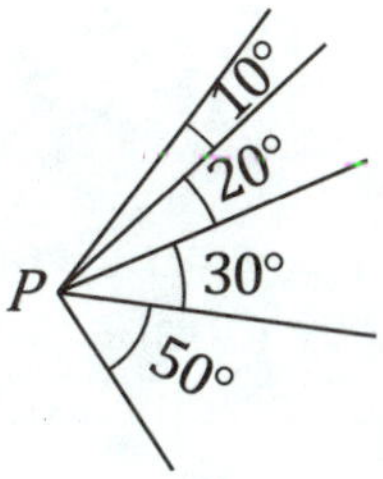

(A) 4 (B) 6 (C) 8 (D) 10 (E) 11

17. How many different three-digit numbers divisible by 25 can be made with the digits 0, 3, 5, and 7 if the digits can be repeated?

(A) 16 (B) 9 (C) 81 (D) 64 (E) 3

18. Each of the boys, Mike, Nate, Oliver, and Paul, has exactly one of the following animals: a cat, a dog, a goldfish, and a canary. Nate has a pet with fur. Oliver has a pet with four legs. Paul has a bird, and Mike and Nate don't like cats. Which of the following sentences is not true?

(A) Oliver has a dog. (B) Paul has a canary. (C) Mike has a goldfish. (D) Oliver has a cat. (E) Nate has a dog.

19. The day after his birthday Johnny said: "The day after tomorrow will be Thursday." On what day of the week did Johnny have his birthday?

(A) on Monday (B) on Tuesday (C) on Wednesday (D) on Thursday (E) on Friday

20. The area of triangle *ABD* is equal to 12, the area of triangle *ABC* is equal to 15 and the area of triangle *ABE* is equal to 4 (see the picture). What is the area of pentagon *ABCED*?

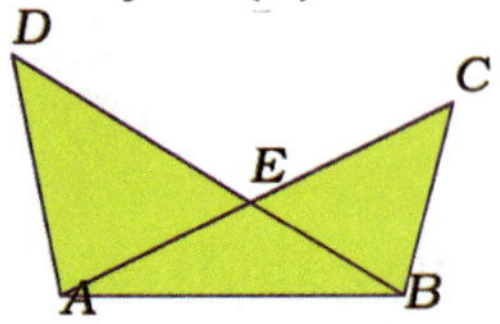

(A) 19 (B) 31 (C) 23 (D) 27 (E) 35

Problems 5 points each

21. The weight of each possible pair of boys from a group of 5 boys was recorded. The following results were obtained: 90 kg, 92 kg, 93 kg, 94 kg, 95 kg, 96 kg, 97 kg, 98 kg, 100 kg, and 101 kg. The total weight of all five boys equals:

(A) 225 kg (B) 230 kg (C) 239 kg (D) 240 kg (E) 250 kg

22. There are four congruent squares. In each of them the midpoints of the sides are indicated and regions with areas S_1, S_2, S_3, and S_4 are shaded (see the picture). Which statement below is true?

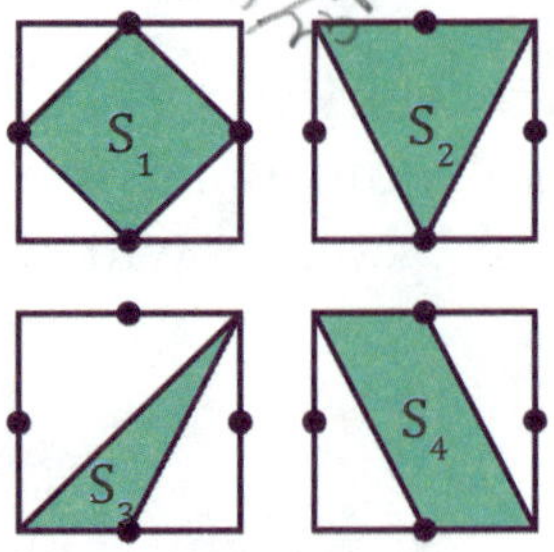

(A) $S_3 < S_4 < S_1 = S_2$ (B) $S_3 < S_1 = S_2 = S_4$ (C) $S_3 < S_1 = S_4 < S_2$ (D) $S_3 < S_4 < S_1 < S_2$ (E) $S_4 < S_3 < S_1 < S_2$

23. You count from 1 to 100 and clap when you say the multiples of the number 3 and when you say numbers that are not multiples of 3 but have 3 as the last digit. How many times will you clap your hands?

(A) 30 (B) 33 (C) 36 (D) 39 (E) 43

24. A cyclist went up the hill with a speed of 12 km/h and went down the hill with a speed of 20 km/h. The ride up the hill took him 16 minutes longer than the ride down the hill. How many minutes did it take the cyclist to go down the hill?

(A) 24 (B) 40 (C) 32 (D) 16 (E) 28

25. The letters P, Q, R, and S indicate the total weight of the figures drawn next to them as shown below.

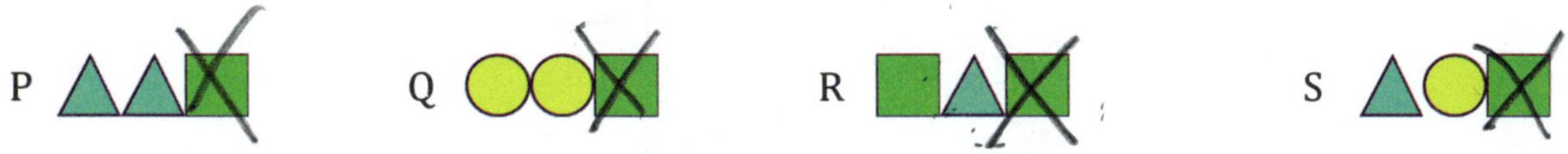

It is known that any two figures of the same shape have the same weight. If $P < Q < R$, then:

(A) $P < S < Q$ (B) $Q < S < R$ (C) $S < P$ (D) $R < S$ (E) $R = S$

26. Ada has 14 gray marbles, 8 white marbles, and 6 black marbles in a bag. What is the least number of marbles she has to take out of her bag with her eyes closed to be sure that she took at least one marble of each color?

(A) 23 (B) 22 (C) 21 (D) 15 (E) 9

27. A certain computer virus destroys computer memory. On the first day it destroyed $\frac{1}{2}$ of the memory of a certain computer. On the second day it destroyed $\frac{1}{3}$ of the memory remaining after the first day. On the third day it destroyed $\frac{1}{4}$ of the memory remaining after the second day, and on the fourth day it destroyed $\frac{1}{5}$ of the memory remaining after the third day. What part of all the computer's memory was left after those four days?

(A) $\frac{1}{5}$ (B) $\frac{1}{6}$ (C) $\frac{1}{10}$ (D) $\frac{1}{12}$ (E) $\frac{1}{24}$

28. What is the greatest value of the sum of the digits of the number made from the sum of the digits of a three-digit number?

(A) 9 (B) 10 (C) 11 (D) 12 (E) 18

29. 32 players were competing in a chess competition. The competition was taking place in stages. In each stage all the players were divided into groups of four. In each of these groups every player played once with each of the other players. Two players from the group went to the next level and the other two players were out of the competition. After the stage in which the four last players played, the two top players played an additional game. How many games were played during the whole competition?

(A) 49 (B) 89 (C) 91 (D) 97 (E) 181

30. A net with 32 hexagonal spaces in three rows was made out of matches as shown below.

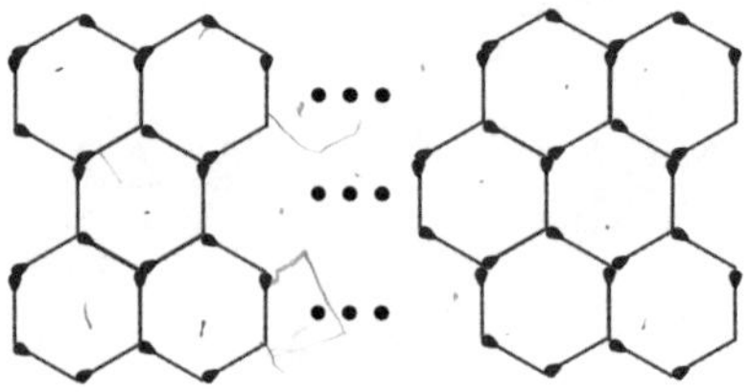

How many matches were used to make this net?

(A) 123 (B) 124 (C) 125 (D) 120 (E) 121

Problems from Year 2004

Problems 3 points each

1. How much is $1000 - 100 + 10 - 1$?

(A) 111 (B) 900 (C) 909 (D) 990 (E) 999

2. In each of the little squares in the table shown to the right, Caroline places one of the digits 1, 2, 3, and 4. She makes sure that each of these numbers is placed in each row and each column. You can see the way she is filling in these squares. What number should she put in the square marked with x?

1		x	2
4	1		
	3		
	2		

(A) 1 (B) 2 (C) 3 (D) 4 (E) It cannot be determined.

3. $(10 \times 100) \times (20 \times 80) =$

(A) $20{,}000 \times 80{,}000$ (B) 2000×8000 (C) $2000 \times 80{,}000$
(D) $20{,}000 \times 8000$ (E) 2000×800

4. 360,000 seconds is:

(A) 3 hours (B) 6 hours (C) 8.5 hours (D) 10 hours (E) more than 90 hours

5. What is the remainder when you divide 20042003 by 2004?

(A) 0 (B) 1 (C) 2 (D) 3 (E) 2003

6. Five identical rectangular plastic sheets were divided into black and transparent squares. Which of the sheets from A to E needs to cover the sheet to the right in order to get a completely black rectangle?

(A)  (B) 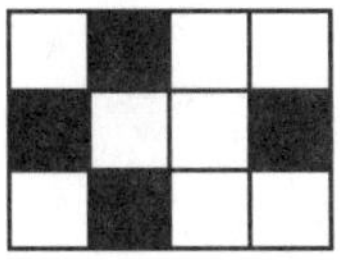 (C) 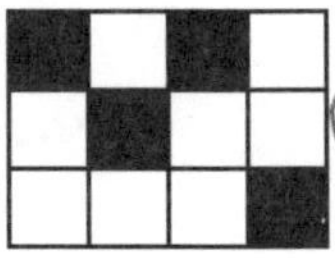 (D)  (E) 

7. Which of the following numbers is not a factor of 2004?

(A) 3 (B) 4 (C) 6 (D) 8 (E) 12

8. Three members of a rabbit family eat 73 carrots altogether during a week. The father ate five carrots more than the mother. Their son ate 12 carrots. How many carrots did the mother eat that week?

(A) 27 (B) 28 (C) 31 (D) 33 (E) 56

9. Nine bus stops are equally spaced along a certain bus route. The distance between the first stop and the third stop is 600 m. How long is the bus route?

(A) 1800 m (B) 2100 m (C) 2400 m (D) 2700 m (E) 3000 m

10. The value of the expression $1 - (2 - (3 - (4 - 5)))$ is equal to:

(A) 0 (B) -3 (C) -9 (D) 3 (E) 9

Problems 4 points each

11. You are given two identical puzzle pieces and you are not allowed to turn them over. Which figure cannot be made out of these two pieces?

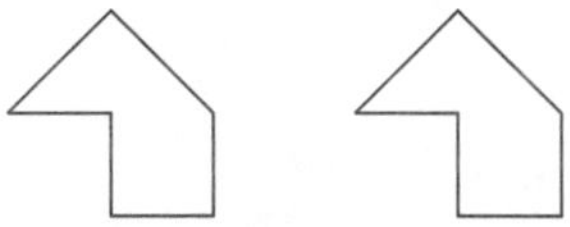

(A) 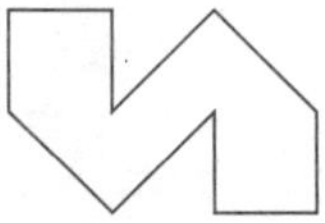 (B) 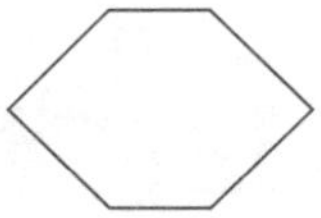 (C) 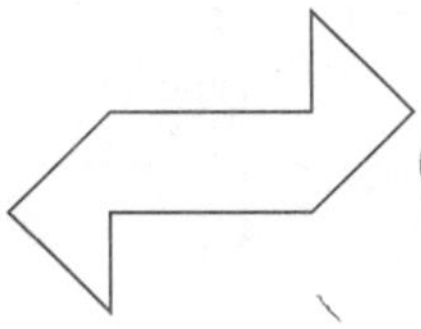 (D) 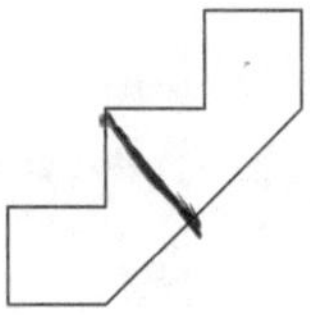 (E) 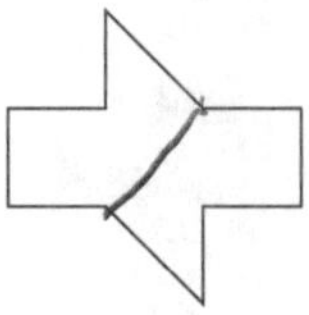

12. Charles folds a sheet of paper in half and then repeats this four more times. Then he makes a hole in the folded paper. How many holes does the sheet of paper have after unfolding?

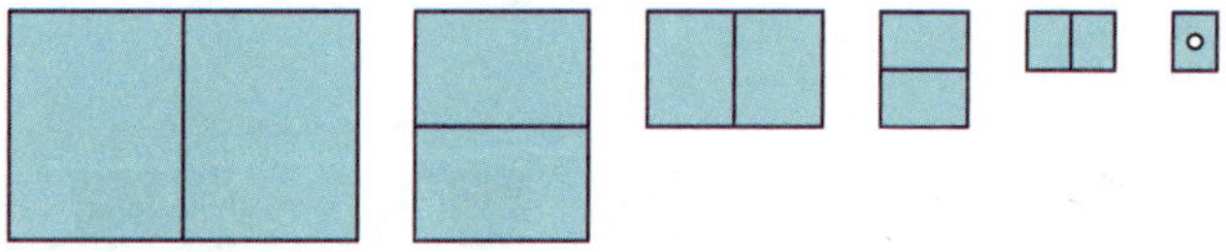

(A) 6 (B) 10 (C) 16 (D) 20 (E) 32

13. Different figures represent different digits and identical figures represent the same digit. Find the digit corresponding to the square.

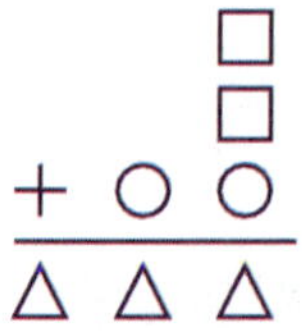

(A) 9 (B) 8 (C) 7 (D) 6 (E) 5

14. The weight of 3 apples and 2 oranges is 255 g. The weight of 2 apples and 3 oranges is 285 g. Each apple weighs the same and each orange weighs the same. What is the combined weight of 1 apple and 1 orange?

(A) 110 g (B) 108 g (C) 105 g (D) 104 g (E) 102 g

15. Thomas, Roman, Andrew, and Michael said the following about a certain number. Thomas said, "This number is equal to 9." Roman said, "This number is prime." Andrew said, "This number is even." Michael said, "This number is equal to 15." Only one of the statements given either by Roman or Thomas is true, and only one of the statements given either by Andrew or Michael is true. What number are they talking about?

(A) 1 (B) 2 (C) 3 (D) 9 (E) 15

16. What is the smallest number of the little squares that have to be shaded in order to get at least one axis of symmetry in the figure shown to the right?

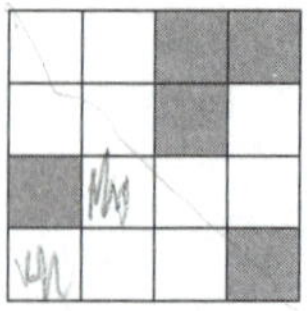

(A) 1 (B) 2 (C) 3 (D) 4 (E) 5

17. One corner of a cube was cut off. Which of the figures below represents the net of the cube after unfolding?

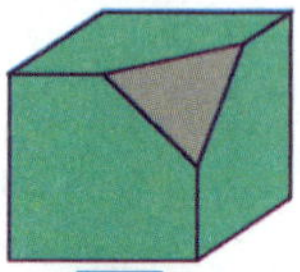

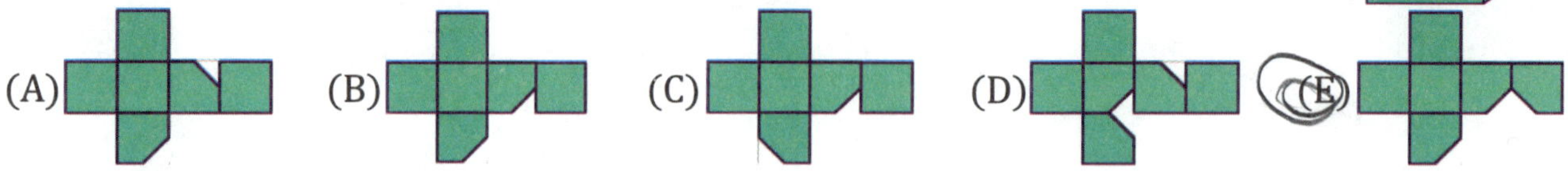

18. Four snails, Fin, Pin, Rin, and Tin, are moving on identical rectangular tiles. The shape and length of each snail's path is shown below. How many decimeters has snail Tin gone?

Snail Fin has gone 25 dm.

Snail Pin has gone 37 dm.

Snail Rin has gone38 dm.

Snail Tin has gone ? dm.

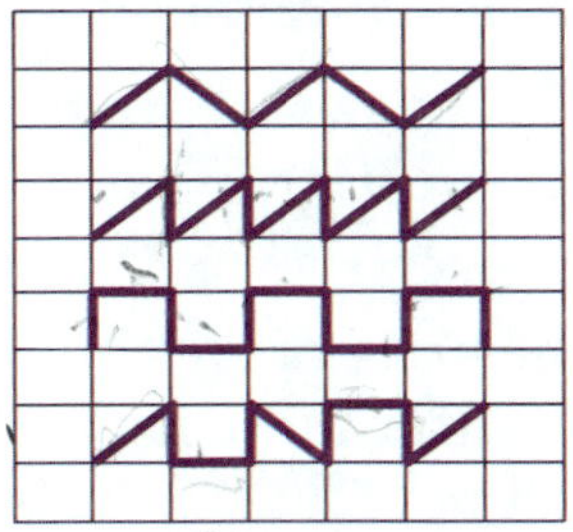

(A) 27 dm (B) 30 dm (C) 35 dm (D) 36 dm (E) 40 dm

19. The Island of Turtles has an unusual weather system: Mondays and Wednesdays are rainy, Saturdays are foggy, and the other days are sunny. A group of tourists would like to go on a 44-day long vacation to the island. Which day of the week should be the first day of their vacation in order to enjoy the most sunny days?

(A) Monday (B) Wednesday (C) Thursday (D) Friday (E) Tuesday

20. The sum of two natural numbers is equal to 77. If the first number is multiplied by 8 and the second by 6, then those products are equal. The larger of these numbers is:

(A) 23 (B) 33 (C) 43 (D) 44 (E) 54

Problems 5 points each

21. The number of all the divisors of the number $2 \times 3 \times 5 \times 7$ is equal to:

(A) 4 (B) 14 (C) 16 (D) 17 (E) 210

22. Together, Ella and Ola have 70 mushrooms. $\frac{5}{9}$ of Ella's mushrooms are brown and $\frac{2}{17}$ of Ola's mushrooms are white. How many mushrooms does Ella have?

(A) 27 (B) 36 (C) 45 (D) 54 (E) 10

23. There are 11 cells shown in the picture:

7								6		

The number 7 is written in the first cell and the number 6 is written in the ninth cell. What number has to be placed in the second cell so that the sum of the numbers from every three consecutive cells is equal to 21?

(A) 7 (B) 8 (C) 6 (D) 10 (E) 21

24. The square shown in the picture was divided into small squares. What fraction of the area of the shaded figure is the area of the figure that is not shaded?

(A) $\frac{1}{4}$ (B) $\frac{1}{5}$ (C) $\frac{1}{6}$ (D) $\frac{2}{5}$ (E) $\frac{2}{7}$

25. In a CD store two CDs had the same price. The price of the first CD was reduced by 5% and the price of the other one was increased by 15%. After this change the prices of the two CDs differ by $6.00. How much does the cheaper CD cost now?

(A) $1.50 (B) $6.00 (C) $28.50 (D) $30.00 (E) $34.50

26. Consecutive natural numbers are placed in the little squares of the big square in the way shown in the figure. Which of the numbers below cannot be placed in the square with the letter x?

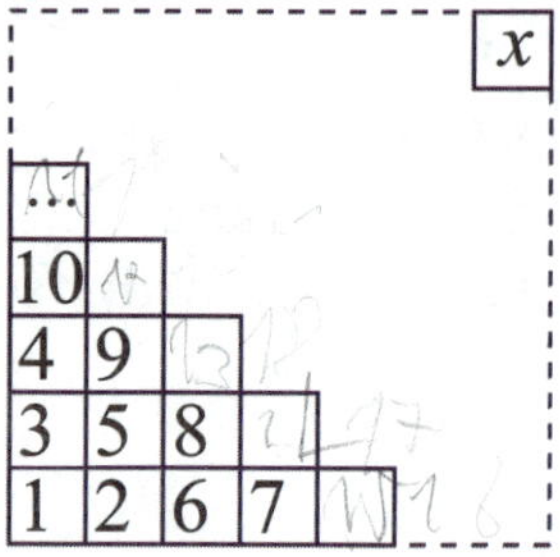

(A) 128 (B) 256 (C) 81
(D) 121 (E) 400

27. Anya divided the number $\underbrace{111\ldots1}_{2004\ times}$ by 3. What is the number of zeros in the quotient?

(A) 670 (B) 669 (C) 668 (D) 667 (E) 665

28. Imagine that you have 108 red marbles and 180 green marbles. The marbles have to be packed in boxes in such a way that every box contains the same number of marbles and there are marbles of only one color in every box. What is the smallest number of boxes that you need?

(A) 288 (B) 36 (C) 18 (D) 8 (E) 1

29. During a competition in the Kangaroo Summer Camp in Zakopane students were given 10 problems to solve. For each correct answer a student was given 5 points and for each incorrect answer the student lost 3 points. Everybody answered all the problems. Mathew got 34 points, Philip got 10 points, and John got 2 points. How many problems did they answer correctly altogether?

(A) 17 (B) 18 (C) 15 (D) 13 (E) 21

30. A right triangle with legs of length 6 cm and 8 cm was cut out of a piece of paper and then folded along a straight line. Which of the numbers below can express the area of the resulting polygon?

(A) 9 cm^2 (B) 12 cm^2 (C) 18 cm^2 (D) 24 cm^2 (E) 30 cm^2

Problems from Year 2006

Problems 3 points each

1. If $3 \times 2006 = 2005 + 2007 + a$, then a is equal to:

(A) 2003 (B) 2004 (C) 2005 (D) 2006 (E) 2007

2. What is the greatest number we can get arranging the six cards with numbers shown in the picture in one row, one after another?

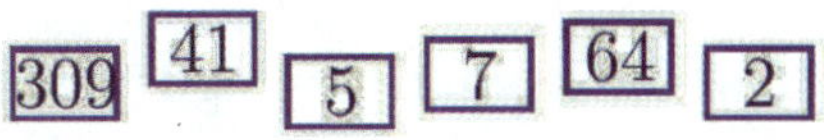

(A) 6,475,413,092 (B) 4,130,975,642 (C) 3,097,564,241 (D) 7,564,413,092 (E) 7,645,413,092

3. There are places for 4 people at a certain square table, one on each side. Students have put together 10 such tables in one row, one next to another, to get one long rectangular table. How many places are there at the rectangular table now?

(A) 40 (B) 32 (C) 30 (D) 22 (E) 20

4. There is an advertisement in a sports store:

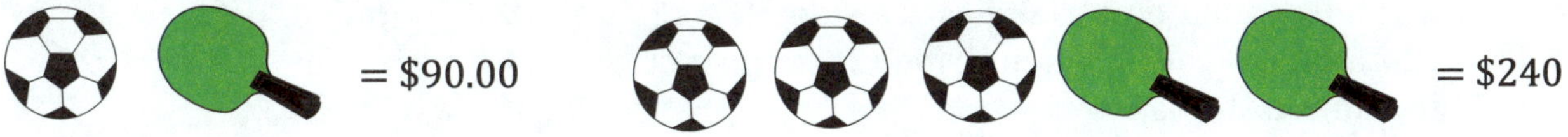

How much does a soccer ball cost?

(A) $130.00 (B) $60.00 (C) $50.00 (D) $40.00 (E) $30.00

5. On which picture do the hands of the clock form an angle with a measure of 150°?

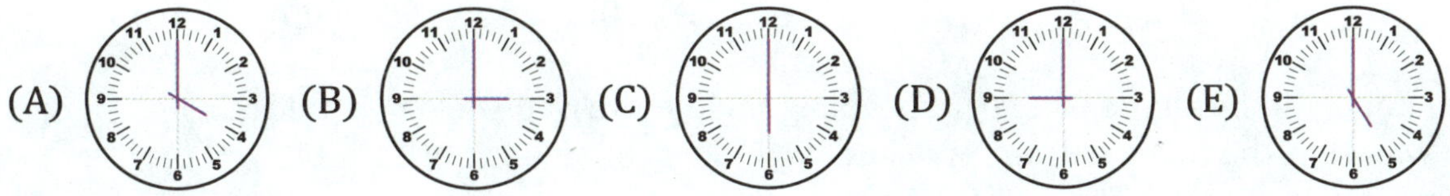

6. On one side of Long Street there are houses numbered by consecutive odd numbers from 1 to 39, and on the other side there are houses numbered by consecutive even numbers from 2 to 34. How many houses are there on Long Street?

(A) 37 (B) 38 (C) 28 (D) 36 (E) 73

7. In how many ways can you get the number 2006 following the arrows in the figure?

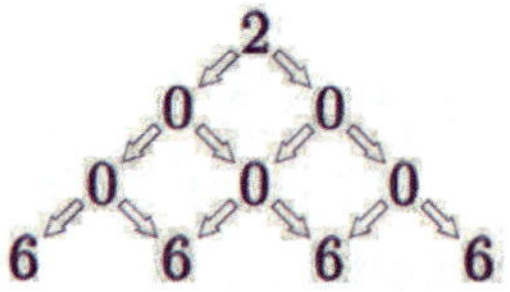

(A) 12 (B) 11 (C) 10 (D) 8 (E) 6

8. One half of one hundredth is:

(A) 0.005 (B) 0.002 (C) 0.05 (D) 0.02 (E) 0.5

9. Out of which of the figures below can you make the box shown in the picture?

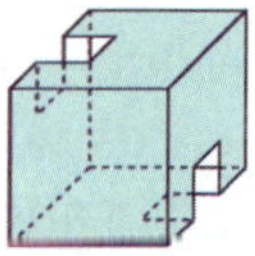

(A) 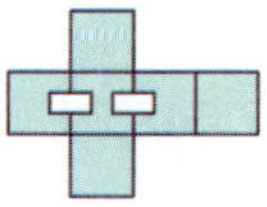 (B) 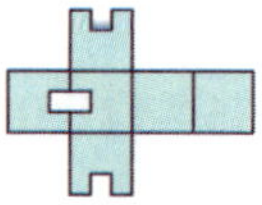 (C) 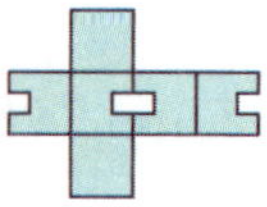 (D) 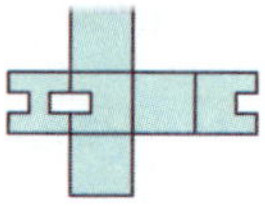 (E) 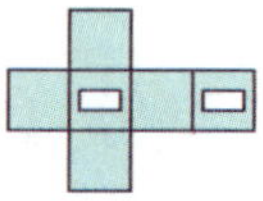

10. The square of the sum of the numbers 5 and 6 decreased by their product equals:

(A) 31 (B) 41 (C) 61 (D) 91 (E) 100

Problems 4 points each

11. The bases of four equilateral triangles are sides of a square in which four circles with radii of 5 have been inscribed (see the picture). The perimeter of the four-pointed star is:

(A) 40 (B) 80 (C) 120 (D) 160 (E) 240

12. What is the difference between the sum of the first 1000 consecutive positive even numbers and the sum of the first 1000 consecutive positive odd numbers?

(A) 1 (B) 1002 (C) 500 (D) 1000 (E) 2000

13. A piece of paper in the shape of a regular hexagon, like the one shown, is folded in such a way that the three corners marked with dots touch each other at the center of the hexagon. The obtained figure is a/an:

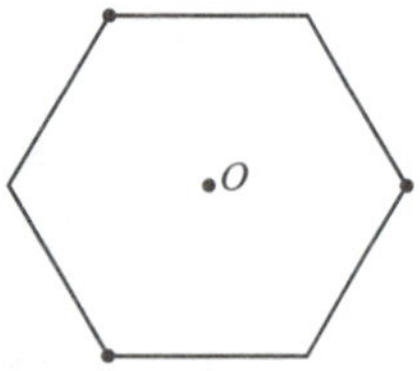

(A) six-pointed star (B) dodecagon (C) hexagon (D) square (E) equilateral triangle

14. To paint all sides of a cube that was built out of little cubes (see Figure 1), 9 fluid ounces of paint were used. How many fluid ounces of paint are needed to paint the white region of the solid shown in Figure 2?

Figure 1

Figure 2

(A) 2 (B) 3 (C) 4.5 (D) 6 (E) 7

15. A car is driving at a constant speed of 25 meters per second. How many kilometers will it travel in one hour? (1 kilometer = 1000 meters)

(A) 100 (B) 90 (C) 80 (D) 75 (E) 60

16. In a rectangle *ABCD*, $|AB| = 4$ inches, and $|BC| = 1$ inch. Point *E* is the midpoint of *AB*, *F* is the midpoint of *AE*, *G* is the midpoint of *AD*, and *H* is the midpoint of *AG*. The area of the shaded rectangle in square inches is:

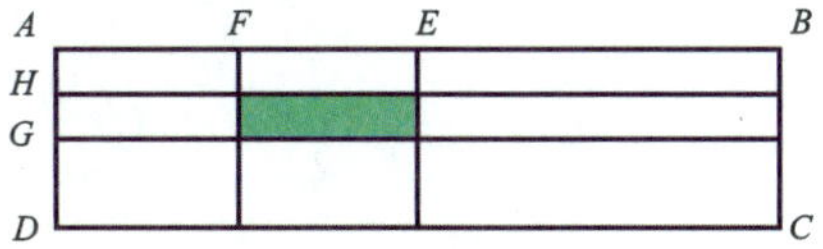

(A) $\frac{1}{4}$ (B) 1 (C) $\frac{1}{8}$ (D) $\frac{1}{2}$ (E) $\frac{1}{16}$

17. What is the result of the addition and subtraction shown to the right?

$$\begin{array}{rr} & 1\,1\,1\,1\,1\,1\,1\,1\,1\,1 \\ - & 1\,1\,1\,1\,1\,1\,1\,1\,1 \\ + & 1\,1\,1\,1\,1\,1\,1\,1 \\ - & 1\,1\,1\,1\,1\,1\,1 \\ + & 1\,1\,1\,1\,1\,1 \\ - & 1\,1\,1\,1\,1 \\ + & 1\,1\,1\,1 \\ - & 1\,1\,1 \\ + & 1\,1 \\ - & 1 \\ \hline \end{array}$$

(A) 111111111 (B) 1010101010 (C) 100000000
(D) 999999999 (E) 1000000000

18. The diameter of the circle in the picture is 10. What is the perimeter of the figure marked with the bold line?

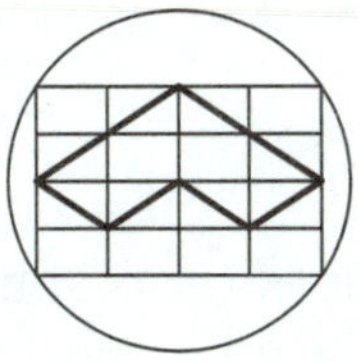

(A) 8 (B) 16 (C) 20 (D) 25 (E) 30

19. Six cars are parked in a parking lot in two rows. Which of the paths from S to F is the shortest?

(A) 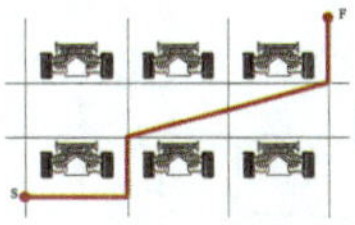 (B) 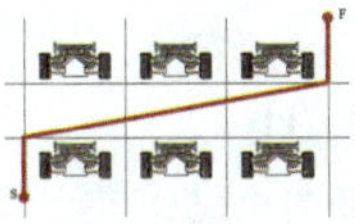 (C) 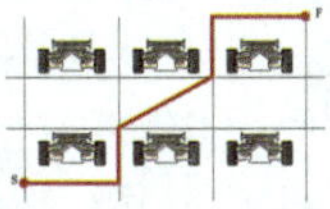 (D) 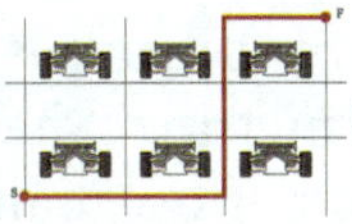 (E) All are equal.

20. Anne added the largest two-digit number divisible by 3 to the smallest two-digit number divisible by 3. Adam added the largest two-digit number not divisible by 3 to the smallest two-digit number not divisible by 3. What is the difference between the sum that Anne calculated and the sum that Adam calculated?

(A) 2 (B) 3 (C) 4 (D) 5 (E) 6

Problems 5 points each

21. On segment $|OE|$ with a length of 2006, we place points A, B, and C so that the length of $|OA| = |BE| = 1111$, and $|OC| = 70\%$ of $|OE|$. What is the order of points A, B, and C on the segment $|OE|$?

(A) A, B, C (B) A, C, B (C) C, B, A (D) B, C, A (E) B, A, C

22. A rope 15 inches long has been divided into the greatest possible number of pieces in such a way that each piece has a different length, and each of these lengths in inches is expressed by a whole number. How many cuts were made?

(A) 3 (B) 4 (C) 5 (D) 6 (E) 7

23. There are two islands on a river that goes through a certain city. There are six connecting bridges as shown in the picture. If we want to travel from point A to point B, starting the journey at bridge 1 and going through each bridge exactly once, then how many possible routes can we take?

(A) 0 (B) 2 (C) 4 (D) 6 (E) more than 6

24. Which set of three numbers represents three points on a number line where one of them is the midpoint of a segment with the ends represented by the other two numbers?

(A) $\frac{1}{3}, \frac{1}{4}, \frac{1}{5}$ (B) 12, 21, 32 (C) $\frac{1}{10}, \frac{9}{80}, \frac{1}{8}$ (D) 0.3, 0.7, 1.3 (E) 24, 48, 64

25. Barbara is creating different squares using sticks of equal length in the way shown in the picture. She labeled the squares with numbers 1, 2, 3, and so on. How many more sticks will she use to create the 31st square compared to the 30th square?

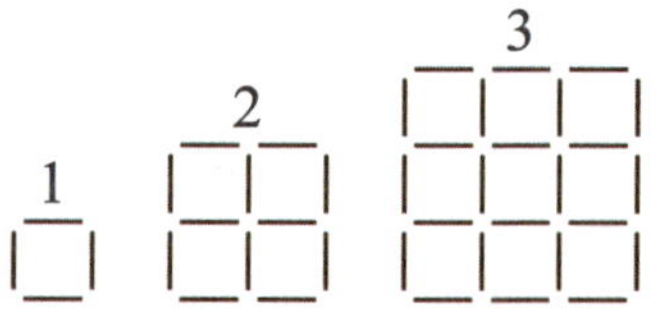

(A) 148 (B) 61 (C) 254 (D) 120 (E) 124

26. Two diagrams of the same cube are shown (see the picture). One letter was written on each side of the cube. In the second figure only two sides have letters on them; the letters on the remaining sides have been erased. What letter was erased from the side marked with the question mark?

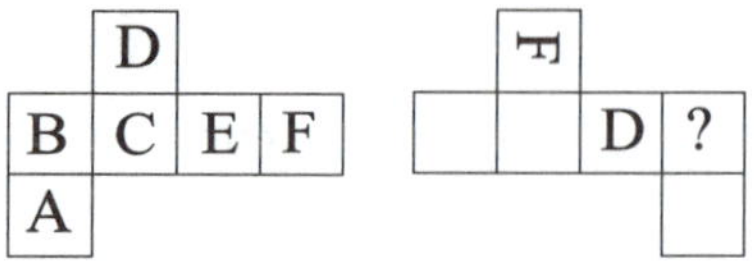

(A) A (B) B (C) C (D) E
(E) It is impossible to determine.

27. A tanker delivered gas to three different gas stations. At the first gas station 30% of the gas was taken out, at the second gas station 40% of the remaining gas in the tank was taken out, and at the third gas station half of the remaining gas was taken out. What percent of the initial amount of gas is left in the tank?

(A) 21 (B) 10 (C) 12 (D) 14 (E) 15

28. In a certain class $\frac{1}{8}$ of the students received a C on the math exam, $\frac{1}{6}$ received a B, and $\frac{2}{3}$ received an A. There were no D's. How many students received an F if there were less than 30 students in the class?

(A) 0 (B) 1 (C) 2 (D) 3 (E) 4

29. Three friends, Adam, Tom, and Paul, went to the swimming pool 15 times. Adam bought the tickets for all of them 8 times and Tom did the same 7 times. Paul gave 30 dollars to his friends, which, as he calculated, he owed for the pool tickets. How should Adam and Tom split those 30 dollars so that each boy pays the same amount for the pool tickets?

(A) \$22 for Adam and \$8 for Tom (B) \$20 for Adam and \$10 for Tom
(C) \$15 for Adam and \$15 for Tom (D) \$16 for Adam and \$14 for Tom
(E) \$18 for Adam and \$12 for Tom

30. All the whole numbers from 1 to 2006 were written on a blackboard. John underlined all the numbers divisible by 2, Adam underlined all the numbers divisible by 3, and Peter underlined all the numbers divisible by 4. How many numbers were underlined exactly twice?

(A) 1003 (B) 668 (C) 501 (D) 334 (E) 167

Problems from Year 2008

Problems 3 points each

1. Which number is the smallest?

(A) $2 + 0 + 0 + 8$ (B) $200 \div 8$ (C) $2 \times 0 \times 0 \times 8$
(D) $200 - 8$ (E) $8 + 0 + 0 - 2$

2. In order to make the expression  ×  $= 2 \times 2 \times 3 \times 3$ true, we need to replace  with:

(A) 2 (B) 3 (C) 2×3 (D) 2×2 (E) 3×3

3. John (J) likes to multiply by 3, Pete (P) likes to add 2, and Nick (N) likes to subtract 1. In what order should they perform their favorite operations if they start with the number 3 and need to end up with 14?

(A) JPN (B) PJN (C) JNP (D) NJP (E) PNJ

4. In order to make the expression $1 + 1♣1 - 2 = 100$ true, we need to replace ♣ with:

(A) + (B) – (C) ÷ (D) 0 (E) 1

5. Carol is playing with two identical cards shaped like equilateral triangles as shown in the picture to the right. She puts the cards either next to each other or partially overlaps them. She then traces the figure made on a piece of paper. There is only one shape below that she cannot get in this way. Which one is it?

(A) 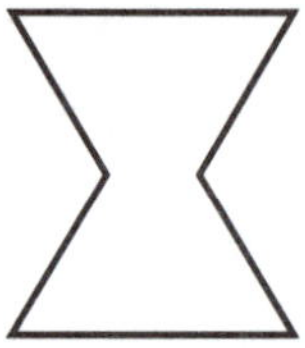 (B) 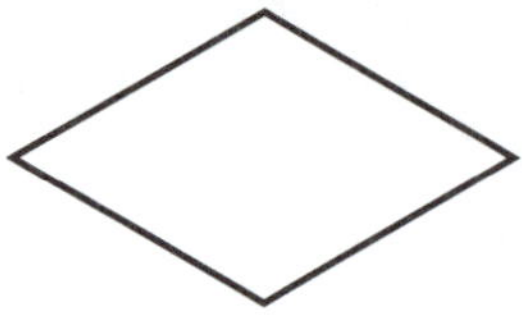 (C) 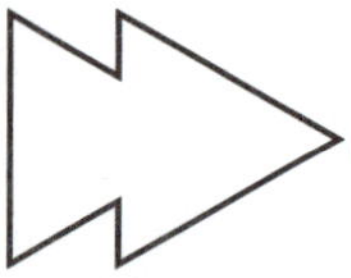 (D)  (E) 

6. At a pirate school, each student had to sew a black and white flag. Exactly three-fifths of the area of the flag had to be black. How many of the flags below were made correctly?

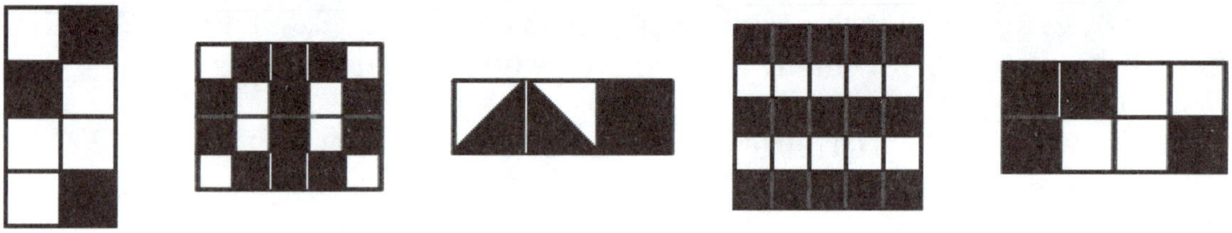

(A) 0 (B) 1 (C) 2 (D) 3 (E) 4

7. Numbers 2, 3, 4 and one more number are written in the cells of 2×2 table. The sum of the numbers in the first row is equal to 9, and the sum of the numbers in the second row is equal to 6. The unknown number is

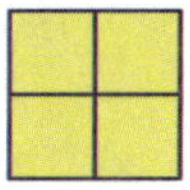

(A) 5 (B) 6 (C) 7 (D) 8 (E) 4

8. Paul had some money in his piggy bank. On his mother's birthday, he borrowed 17 dollars from his sister and bought a present for his mother for 21 dollars. He had 15 dollars left over. How many dollars did Paul have in his piggy bank in the beginning?

(A) 32 (B) 11 (C) 53 (D) 38 (E) 19

9. The first picture to the right shows a multiplication table. The second also shows a multiplication table, but some of the numbers have been erased. What number was found in the square with the question mark?

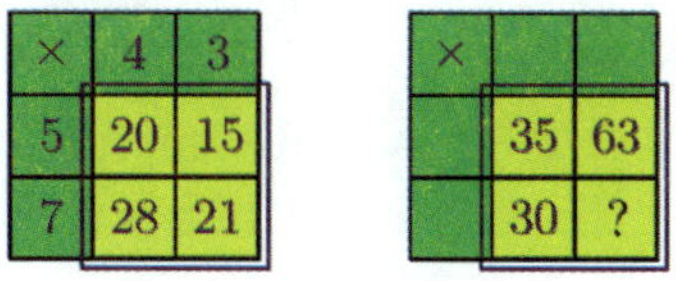

(A) 54 (B) 56 (C) 65 (D) 36 (E) 42

10. A toy store was selling a brick tower, which is a figure made of black and white bricks as shown in Figure 1. Each layer of the figure is made up of bricks of the same color. Figure 2 shows a view of this tower from the top. How many white bricks were used to make the tower?

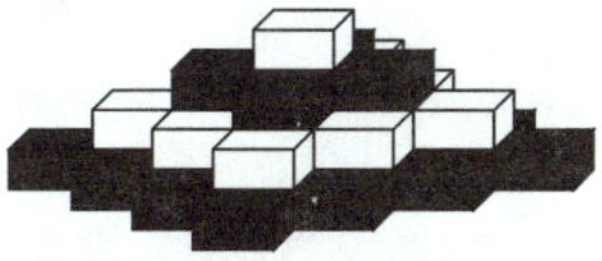
Figure 1

Figure 2

(A) 9 (B) 10 (C) 12 (D) 13 (E) 14

Problems 4 points each

11. There are five boxes as shown in the picture, and each one contains cards with different letters. Paul wants to remove cards from the boxes in such a way that there is only one card left in each box, and that every box has a card with a different letter in it. Which card will be left in box 5?

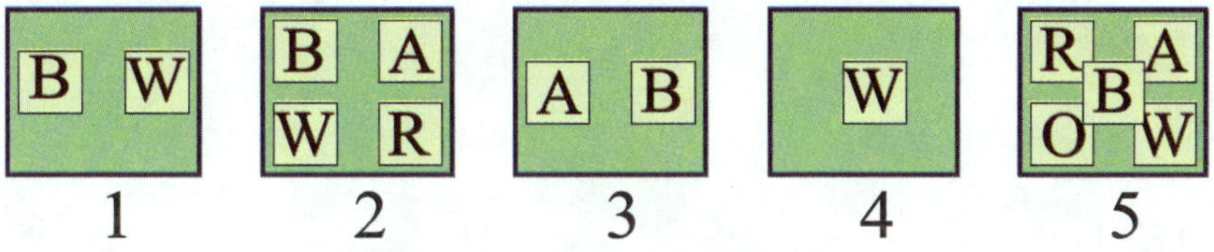

(A) B (B) R (C) A (D) W (E) O

12. In the picture to the right, the perimeters of the square and of the triangle are equal. What is the perimeter of the pentagon that they compose when put together in the way shown?

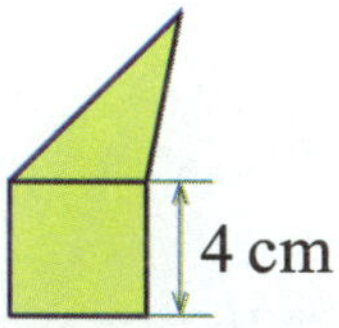

(A) 12 cm (B) 24 cm (C) 28 cm (D) 32 cm (E) 18 cm

13. Given the choice of numbers listed below, using how many identical matches can you not make a triangle? (You cannot break the matches!)

(A) 7 (B) 6 (C) 5 (D) 4 (E) 3

14. What is the area of a square that has a side 5 cm longer than the side of a square with an area of 121 cm^2?

(A) 196 cm^2 (B) 400 cm^2 (C) 324 cm^2 (D) 289 cm^2 (E) 256 cm^2

15. Mark had a full bottle of juice. He poured $\frac{1}{3}$ of what was in the bottle into a glass, and then $\frac{3}{4}$ of what was left in the bottle into a pitcher. What fraction of the original amount of juice was left in the bottle?

(A) $\frac{1}{4}$ (B) $\frac{2}{3}$ (C) $\frac{11}{12}$ (D) $\frac{1}{6}$ (E) This cannot be determined.

16. If we throw a dart at a dartboard, we can get 2, 3, or 6 points if we hit the target (see the picture), or 0 points if we miss. How many different scores can we obtain by throwing two darts?

(A) 4 (B) 6 (C) 8 (D) 9 (E) 10

17. Alexa wanted to put all her CDs on a shelf, but one third of them did not fit. She put some of the CDs that did not fit on the shelf into three boxes. She put seven CDs in each box, and she still had two CDs that did not fit. She left these last two CDs on her desk. How many CDs does Alexa have?

(A) 23 (B) 81 (C) 69 (D) 67 (E) 93

18. Each of the figures (A) to (E) shown below is made of exactly 5 blocks. Which of them can you not make if you start with the figure on the right and are allowed to move only one block?

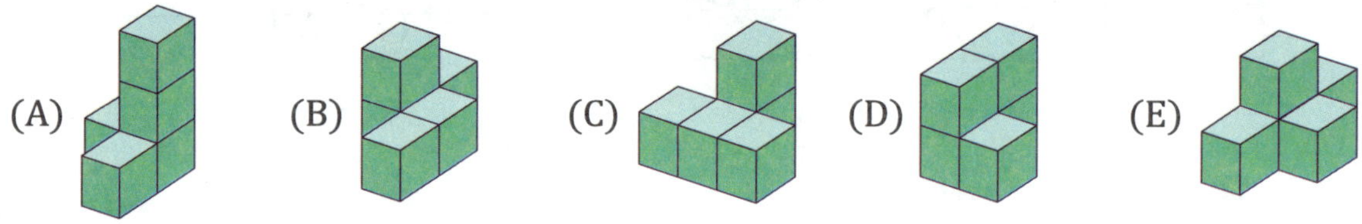

19. The sum of the digits of the result of the operation $10^{101} - 9$ is

(A) 891 (B) 901 (C) 991 (D) 10 (E) 900

20. In two years, the son of Mr. and Mrs. Kowalski will be two times as old as he was two years ago. In three years, their daughter will be three times as old as she was three years ago. Which of the statements below is true?

(A) The son is one year older than the daughter.
(B) The daughter is one year older than the son.
(C) The son and the daughter are the same age.
(D) The son is two years older than the daughter.
(E) The daughter is two years older than the son.

Problems 5 points each

21. The five symbols used in the operations below represent different digits.

@ + @ + @ = *

+ # + # = &

* + & = ^

What digit is represented by ^?

(A) 0 (B) 2 (C) 6 (D) 8 (E) 9

22. Three friends, Smith, Roberts, and Farrell, each have one of three professions: doctor, engineer, and musician. Each one has a different profession. The doctor does not have a sister nor a brother. He is the youngest of the three friends. Farrell is older than the engineer and is married to Smith's sister. The names of the doctor, the engineer, and the musician are listed in the following order:

(A) Farrell, Roberts, Smith (B) Farrell, Smith, Roberts (C) Roberts, Smith, Farrell
(D) Roberts, Farrell, Smith (E) Smith, Farrell, Roberts

23. A certain robot is moving through a small, checkered board (see the picture). In each move, it can move to a neighboring square, that is, to a square that shares a side with the square the robot is already on. The robot has to move through each of the squares on the board exactly once. To do this, it can start:

(A) on the middle square only
(B) on one of the corner squares only
(C) on any of the non-shaded squares
(D) on any of the shaded squares
(E) on any of the squares

24. The picture shows the plan of a town. There are four bus routes in the town. Bus L1 follows the route C-D-E-F-G-H-C, which is 17 km long. Bus L2 follows the route A-B-C-F-G-H-A, which is 12 km long. The route of bus L3 is A-B-C-D-E-F-G-H-A, which is 20 km long. Bus L4 travels along the route C-F-G-H-C. How long is the route of bus L4?

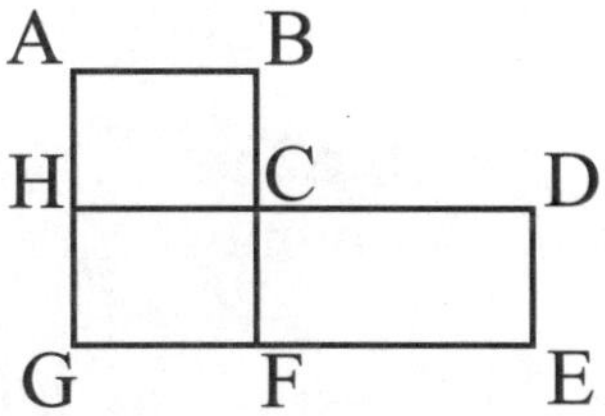

(A) 5 km (B) 8 km (C) 9 km (D) 12 km (E) 15 km

25. Points *A*, *B*, *C*, and *D* are marked on a straight line in a certain order. We know that $|AB| = 13$, $|BC| = 11$, $|CD| = 14$, and $|DA| = 12$. What is the distance between the two points farthest from each other?

(A) 14 (B) 38 (C) 50 (D) 25 (E) 23

26. A train traveling at a steady speed crossed a bridge which was 200 meters long in 1 minute. The whole train passed a person standing on the bridge in 12 seconds. How long was the train?

(A) 100 meters (B) 60 meters (C) 50 meters (D) 40 meters (E) 75 meters

27. A fairy has 6 bottles. Their volumes are 16 oz, 18 oz, 22 oz, 24 oz, 32 oz, and 34 oz. Some are filled with orange juice, some are filled with cherry juice, and one is empty. There is twice as much orange juice as cherry juice. What is the volume of the empty bottle?

(A) 18 oz (B) 34 oz (C) 24 oz (D) 32 oz (E) 22 oz

28. In every two-digit number, the digit in the ones place was subtracted from the digit in the tens place. The sum of all the results is:

(A) 90 (B) 100 (C) 55 (D) 45 (E) 30

29. How many natural numbers are there for which the quotient $\frac{n+41}{n+5}$ is a natural number?

(A) 1 (B) 2 (C) 4 (D) 5 (E) 36

30. What is the largest number of digits that can be erased from the 1000-digit number 20082008…2008 so that the sum of the remaining digits is 2008?

(A) 260 (B) 510 (C) 746 (D) 1020 (E) 130

Problems from Year 2010

Problems 3 points each

1. Each ▲ in the equation ▲ + ▲ + 6 = ▲ + ▲ + ▲ + ▲ represents the same number. This number is

(A) 2 (B) 3 (C) 4 (D) 5 (E) 6

2. Which of the following numbers is prime?

(A) $201+0$ (B) $2+0-1+0$ (C) 20×10
(D) $2+0+1+0$ (E) $20\times 1\times 0$

3. Ella walked directly from home to school. She did not stop and she did not turn back. Which of the following numbers definitely cannot be the number of flowers Ella passed on her way to school?

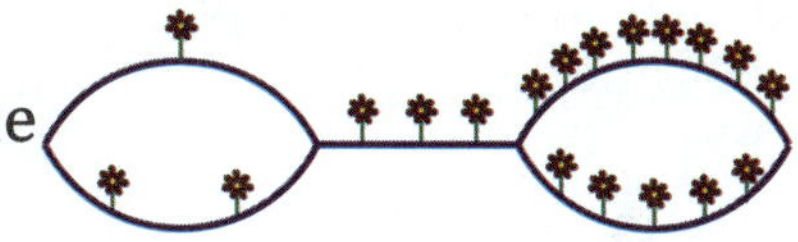

(A) 9 (B) 10 (C) 11 (D) 12 (E) 13

4. Anna connected some points with line segments as shown. How many segments did Anna draw?

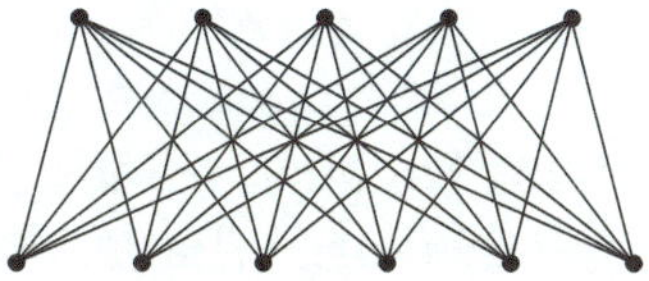

(A) 20 (B) 25 (C) 30 (D) 40 (E) 60

5. Two years ago the sum of the ages of two cats, Mr. Meow and Mr. Whiskers, was 15. Now Mr. Meow is 13 years old. In how many years will Mr. Whiskers be 9 years old?

(A) in 1 year (B) in 2 years (C) in 3 years (D) in 4 years (E) in 5 years

6. A square piece of paper is white on one side and green on the other. Eve divided it into 9 little squares. She labeled some edges with natural numbers 1 to 8 (see Figure 1). Along which of these numbered edges does she need to make cuts to obtain the shape shown in Figure 2?

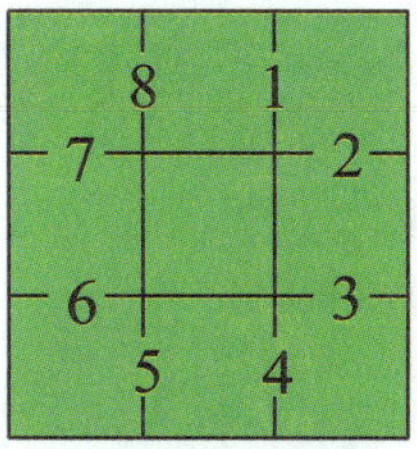

Figure 1

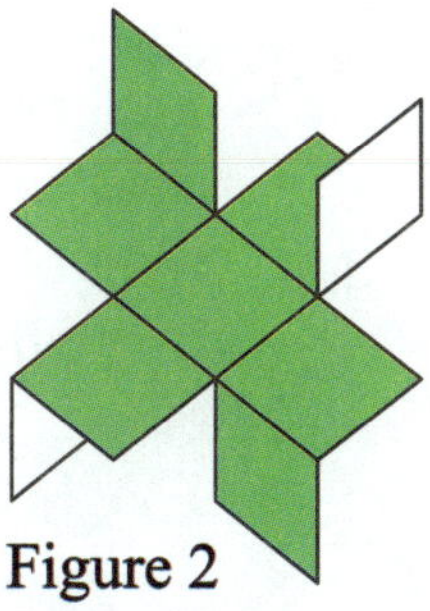
Figure 2

(A) 1, 3, 5 and 7 (B) 2, 4, 6 and 8 (C) 2, 3, 5 and 6
(D) 2, 4, 6 and 7 (E) 1, 4, 5 and 8

7. Seven identical candy bars were arranged on the bottom of a square box. It is possible to slide the candy bars so that there will be room for one more candy bar. What is the least number of candy bars that must be moved to make room for one more candy bar?

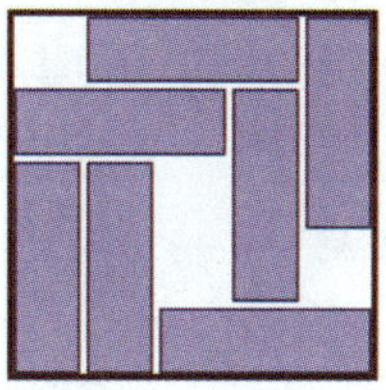

(A) 1 (B) 2 (C) 3 (D) 4 (E) 5

8. An old book was missing several consecutive pages. The last page number before the missing pages was 28 and the first page number after the missing pages was 75. How many sheets were missing from the old book?

(A) 51 (B) 23 (C) 22 (D) 21 (E) 50

9. Which of the following expressions has a different value from all the others?

(A) $20 \times 10 + 20 \times 10$ (B) $20 \div 10 \times 20 \times 10$ (C) $20 \times 10 \times 20 \div 10$
(D) $20 \times 10 + 10 \times 20$ (E) $20 \div 10 \times 20 + 10$

10. A fly has 6 legs and a spider has 8 legs. Together, 2 flies and 3 spiders have as many legs as 10 birds and

(A) 2 cats (B) 3 cats (C) 4 cats (D) 5 cats (E) 6 cats

Problems 4 points each

11. Each pair of adjacent sides of the figure shown in the picture to the right is perpendicular. What is the perimeter of this figure?

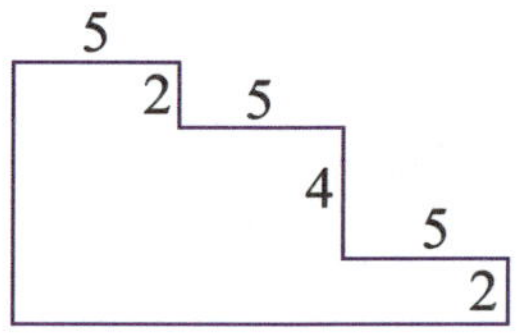

(A) $3 \times 5 + 4 \times 2$ (B) $3 \times 5 + 8 \times 2$ (C) $6 \times 5 + 4 \times 2$
(D) $6 \times 5 + 6 \times 2$ (E) $6 \times 5 + 8 \times 2$

12. Adam divided a certain secret number by 7. Then, he added 7 to the result and multiplied the number he obtained in this way by 7. His final result was 777. What was Adam's secret number?

(A) 770 (B) 111 (C) 722 (D) 567 (E) 728

13. Angle ABC of the quadrilateral $ABCD$ shown in the picture to the right measures

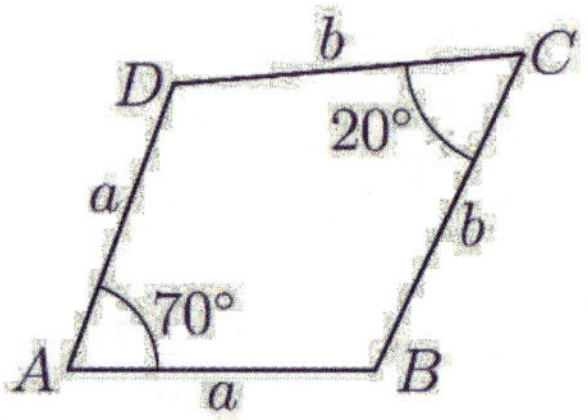

(A) 110° (B) 120° (C) 125° (D) 135° (E) 140°

14. Rectangle $ABCD$ has a perimeter of 120 cm and its diagonals intersect at point P. The distance from P to the side BC is twice the distance from P to side AB. The area of the rectangle equals

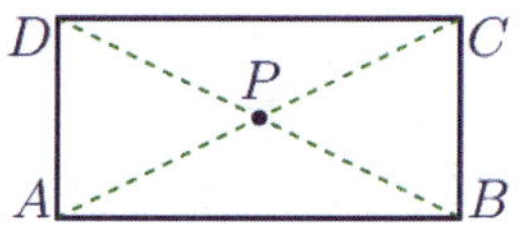

(A) 200 cm^2 (B) 400 cm^2 (C) 600 cm^2 (D) 800 cm^2 (E) 1000 cm^2

15. An ant is walking along the lines of the grid (see the picture to the right). It always begins and ends its walk at point A but cannot visit any other point more than once. The ant must walk over the segments indicated in bold. The least number of squares enclosed by the ant's path is

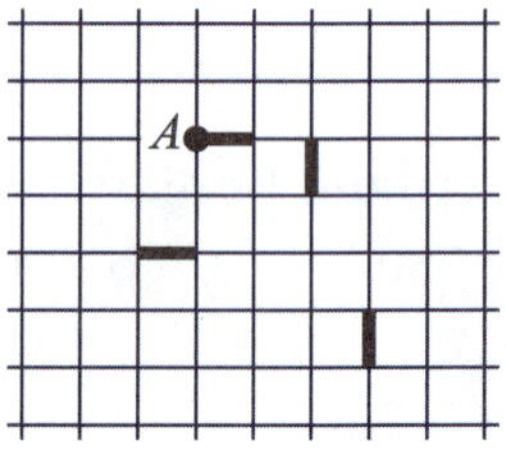

(A) 11 (B) 9 (C) 7 (D) 8 (E) 13

16. Sophie painted a 4 on a square. She folded the square in half, unfolded it, and folded it in half again. Figure 1 shows the result. She did likewise with the number 5. Which of the pictures below has the same image as the small square indicated with the question mark in Figure 2?

Figure 1

Figure 2

(A) 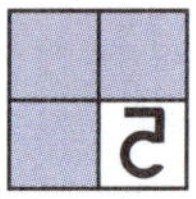 (B) 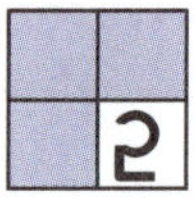 (C)  (D) 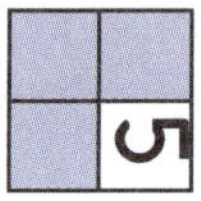 (E) 

17. What fraction of the square is the shaded region?

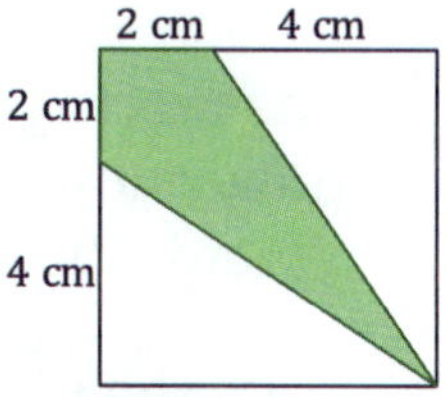

(A) $\frac{1}{3}$ (B) $\frac{1}{4}$ (C) $\frac{1}{5}$ (D) $\frac{3}{8}$ (E) $\frac{2}{9}$

18. How many black squares must be painted white so that each row and each column of the grid contains only one black square?

(A) 4 (B) 5 (C) 6 (D) 7
(E) This is not possible.

19. The figure to the right shows three identical dice that have been glued together. The sum of the number of dots on opposite faces of every die is always 7. What is the sum of the number of dots on the faces that have been glued together?

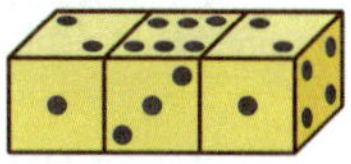

(A) 12 (B) 13 (C) 14 (D) 15 (E) 16

20. Numbers 1, 4, 7, 10, and 13 were entered into the grid shown to the right in such a way that the sum of the numbers in the row equals the sum of the numbers in the column. What is the largest possible sum of the numbers in the row obtained in this way?

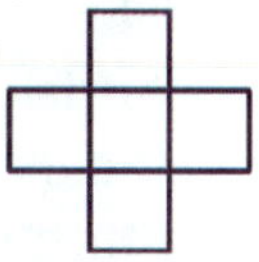

(A) 18 (B) 27 (C) 21 (D) 30 (E) 24

Problems 5 points each

21. A pizza parlor sells small, medium, and large pizzas. Each pizza is made with cheese, tomatoes, and at least one of the following toppings: pepperoni, ham, mushrooms, and olives. How many different pizzas are possible?

(A) 15 (B) 30 (C) 12 (D) 45 (E) 48

22. A jeweler can make chains of any length using identical links. Figure 1 shows such a chain made of three links. A single link is shown in Figure 2. What is the length of a chain made of seven links?

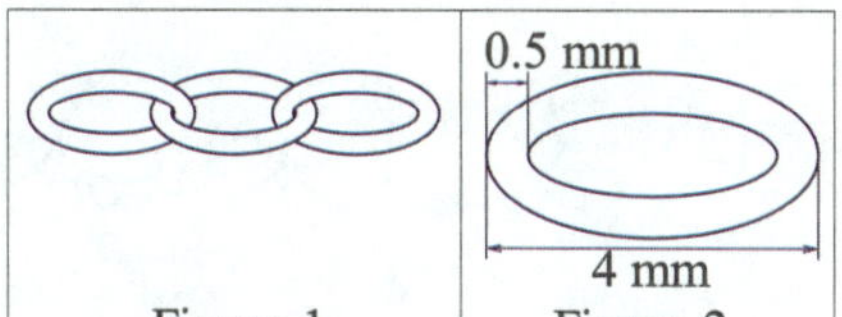

Figure 1 Figure 2

(A) 28 mm (B) 23 mm (C) 22 mm (D) 22.5 mm (E) 21 mm

23. Adam and Tom are walking in the same direction around a circular table and counting chairs. They begin their counts with different chairs. Tom's twentieth chair is Adam's fourth chair, while Tom's tenth chair is Adam's forty-sixth chair. How many chairs are there at the table?

(A) 50 (B) 52 (C) 56 (D) 60 (E) 80

24. The rectangle shown in the picture was divided into squares with different side lengths. The areas of some of these squares are given. What is the area of the entire rectangle?

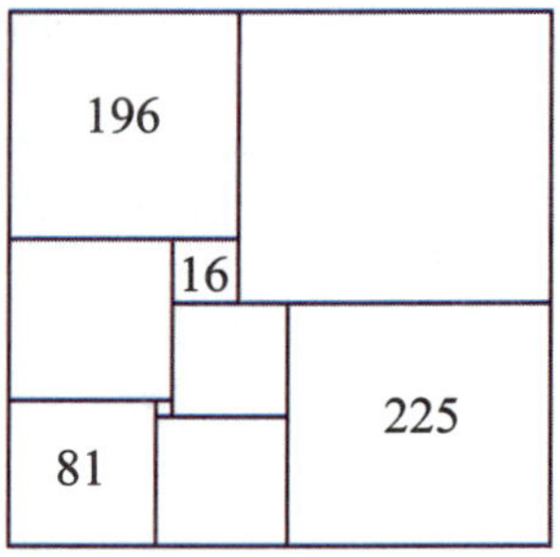

(A) 1024 (B) 1056 (C) 1089
(D) 1120 (E) 1122

25. How many numbers are there that have the sum of their digits equal to 100 and the product of their digits equal to 5?

(A) 2010 (B) 105 (C) 100 (D) 95 (E) 96

26. How many five-digit numbers in the form 1_82_ are there that are divisible by 12 and all of whose digits are different?

(A) 8 (B) 6 (C) 10 (D) 4 (E) 2

27. Theater seats are numbered as shown in the picture. Anna and Eve were standing in line to buy tickets for the evening's performance. Anna bought a ticket for seat number 100. At this time there were only 5 tickets left. These were for the seats numbered 76, 94, 99, 104, and 118. Eve, who bought her ticket right after Anna, bought it for the seat closest to her. For what number seat was Eve's ticket?

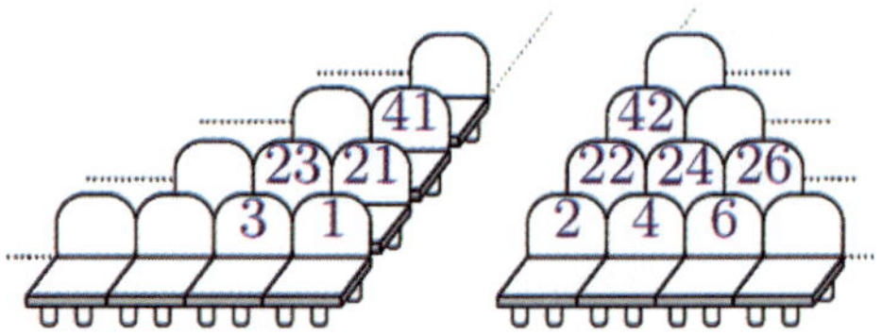

(A) 94 (B) 76 (C) 99 (D) 104 (E) 118

28. Wally wrapped a wire around a notched board. The picture to the right shows the front side of the board. Which of the pictures below shows the back side of this board?

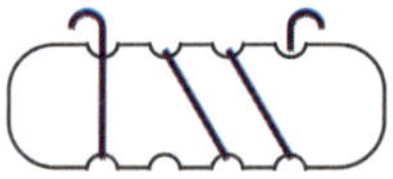

(A)  (B)  (C) 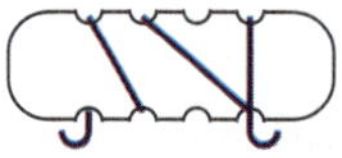 (D) 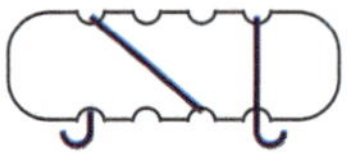 (E) 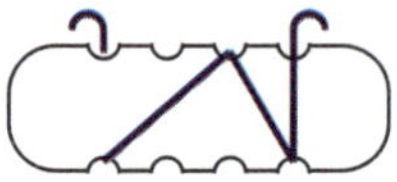

29. Consider all natural numbers m and n, both different than 0, that satisfy the condition $75 \times m = n^3$. The smallest possible value of the sum $m + n$ is

(A) 15 (B) 30 (C) 50 (D) 60 (E) 5700

30. A magical kingdom is inhabited by dragons with six, seven, and eight heads. Those with 7 heads always lie, and those with 6 or 8 heads always tell the truth. One day four dragons met. The blue dragon said, "Together we have 28 heads," the green dragon said, "Together we have 27 heads," the yellow dragon said, "Together we have 26 heads," and the red dragon said, "Together we have 25 heads." What color was the dragon that did not lie?

(A) red (B) blue (C) green (D) yellow (E) It is impossible to determine.

Problems from Year 2012

Problems 3 points each

1. Basil wants to paint the slogan VIVAT KANGAROO on a wall. He wants to paint different letters different colors and the same letters the same color. How many colors will he need?

(A) 7 (B) 8 (C) 9 (D) 10 (E) 13

2. A blackboard is 6 m wide. The width of the middle part is 3 m. The two other parts have equal widths. How wide is the right-hand part?

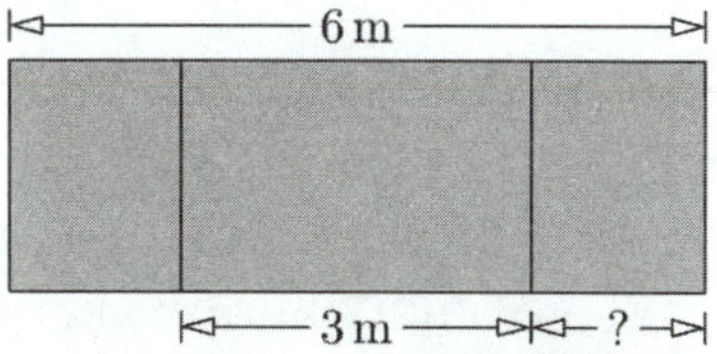

(A) 1 m (B) 1.25 m (C) 1.5 m (D) 1.75 m (E) 2 m

3. Sally can put 4 coins in a square made using 4 matches (see picture). At least how many matches will she need in order to make a square containing 16 coins that do not overlap?

(A) 8 (B) 10 (C) 12 (D) 15 (E) 16

4. On a certain plane, the rows are numbered from 1 to 25, but there is no row number 13. Row number 15 has only 4 passenger seats; all the other rows have 6 passenger seats. How many seats for passengers are there on this plane?

(A) 120 (B) 138 (C) 142 (D) 144 (E) 150

5. When it is 4 o'clock in the afternoon in London, it is 5 o'clock in the afternoon in Madrid and it is 8 o'clock in the morning on the same day in San Francisco. Ann went to bed in San Francisco at 9 o'clock yesterday evening. What was the time in Madrid at that moment?

(A) 6 o'clock yesterday morning
(B) 6 o'clock yesterday evening
(C) 12 o'clock yesterday afternoon
(D) 12 o'clock midnight
(E) 6 o'clock this morning

6. The picture to the right shows a pattern of hexagons. We draw a new pattern by connecting all the centers of neighboring hexagons. Which of the figures below do we get?

(A) 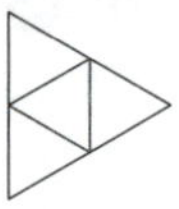 (B)  (C) 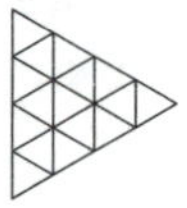 (D)  (E) 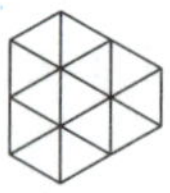

7. We add 3 to 6. Then we multiply the result by 2 and then we add 1. Then the final result will be the same as the result of the computation

(A) $(6 + 3 \times 2) + 1$ (B) $6 + 3 \times 2 + 1$ (C) $(6 + 3) \times (2 + 1)$
(D) $(6 + 3) \times 2 + 1$ (E) $6 + 3 \times (2 + 1)$

8. The upper coin is rotated without slipping around the fixed lower coin to a position shown in the picture. Which is the resulting relative position of the kangaroos?

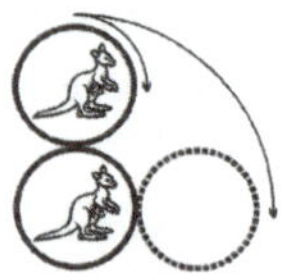

(A)  (B)  (C)  (D) 
(E) It depends on the rotation speed.

9. One balloon can lift a basket containing items weighing at most 80 kg. Two such balloons can lift the same basket containing items weighing at most 180 kg. What is the weight of the basket?

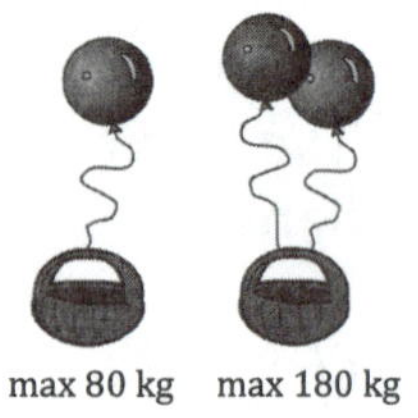

(A) 10 kg (B) 20 kg (C) 30 kg (D) 40 kg (E) 50 kg

10. Vivien and Mike were given some apples and pears by their grandmother. Altogether, they had 25 pieces of fruit in their baskets. On the way home, Vivien ate 1 apple and 3 pears and Mike ate 3 apples and 2 pears. At home they found out that they brought home the same number of pears as apples. How many pears were they given by their grandmother?

(A) 12 (B) 13 (C) 16 (D) 20 (E) 21

Problems 4 points each

11. Which three of the numbered puzzle pieces should you add to the picture to complete the square?

(A) 1, 3, 4 (B) 1, 3, 6 (C) 2, 3, 5 (D) 2, 3, 6 (E) 2, 5, 6

12. Lisa had 8 dice with the letters A, B, C, and D, with the same letter on all sides of each die. She builds a block with them. Two adjacent dice always have different letters. What letter is on the die that cannot be seen in the picture?

(A) A (B) B (C) C (D) D (E) It is impossible to say.

13. There are five cities in Wonderland. Each pair of cities is connected by one road, either visible or invisible. On the map of Wonderland, there are only seven visible roads, as shown. Alice has magical glasses: when she looks at the map through these glasses, she only sees the roads that are otherwise invisible. How many invisible roads can she see?

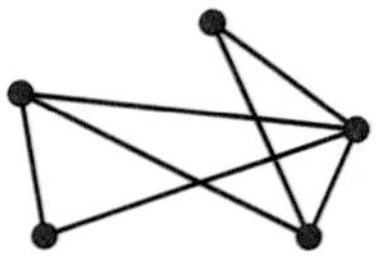

(A) 9 (B) 8 (C) 7 (D) 3 (E) 2

14. The positive integers have been colored red, blue or green: 1 is red, 2 is blue, 3 is green, 4 is red, 5 is blue, 6 is green, and so on. Renate calculates the sum of a red number and a blue number. What color can the resulting number be?

(A) It is impossible to say. (B) red or blue (C) only green
(D) only red (E) only blue

15. The perimeter of the figure to the right, made up of identical squares, is equal to 42 cm. What is the area of the figure?

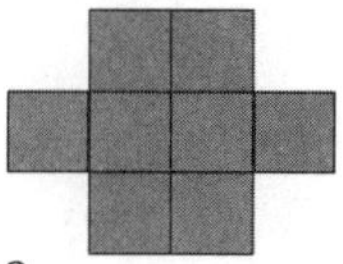

(A) 8 cm^2 (B) 9 cm^2 (C) 24 cm^2 (D) 72 cm^2 (E) 128 cm^2

16. Look at the pictures. Both shapes are formed from the same five pieces. The rectangle measures 5 cm × 10 cm, and the other parts are quarters of two different circles. The difference between the lengths of the perimeters of the two shapes is

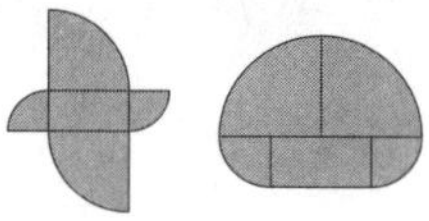

(A) 2.5 cm (B) 5 cm (C) 10 cm (D) 20 cm (E) 30 cm

17. Place the numbers from 1 to 7 in the circles in such a way that the sum of the numbers on each of the indicated lines of three circles is the same. What is the number at the top of the triangle?

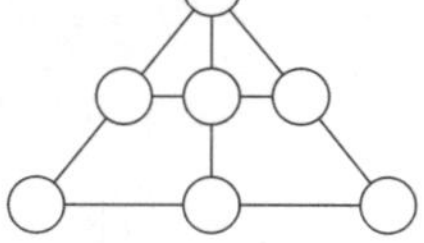

(A) 1 (B) 3 (C) 4 (D) 5 (E) 6

18. A rubber ball falls vertically from a height of 10 m from the roof of the house. After each impact on the ground it bounces back up to ⅘ of the previous height. How many times will the ball appear in front of a rectangular window whose bottom edge is at a height of 5 m and whose top edge is at a height of 6 m?

(A) 3 (B) 4 (C) 5 (D) 6 (E) 8

19. There are 4 gears on fixed axles next to each other, as shown. The first one has 30 teeth, the second one 15, the third one 60 and the last one 10. How many revolutions does the last gear wheel make when the first one turns through one revolution?

(A) 3 (B) 4 (C) 6 (D) 8 (E) 9

20. A regular octagon is folded in half exactly three times until a triangle is obtained, as shown. Then the apex is cut off at a right angle, as shown in the picture. When the paper is unfolded, what will it look like?

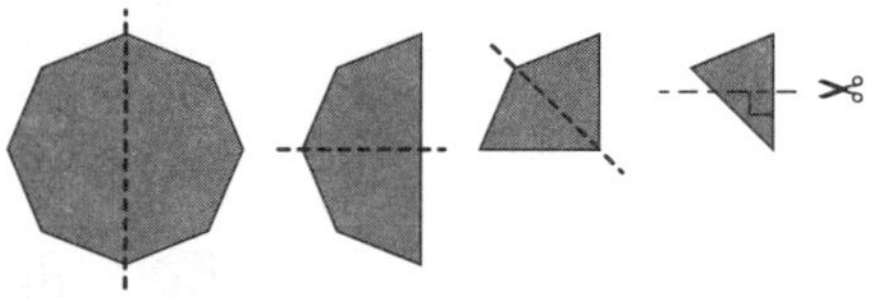

(A)  (B) 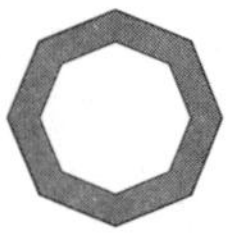 (C) 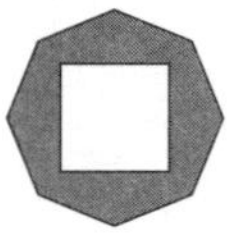 (D)  (E) 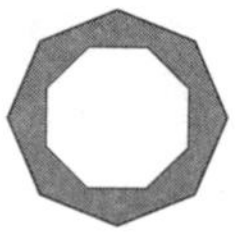

Problems 5 points each

21. Winnie's vinegar-wine-water marinade contains vinegar and wine in the ratio of 1 to 2, and wine and water in the ratio of 3 to 1. Which of the following statements is true?

(A) There is more vinegar than wine.
(B) There is more wine than vinegar and water together.
(C) There is more vinegar than wine and water together.
(D) There is more water than vinegar and wine together.
(E) There is less vinegar than either water or wine.

22. Kangaroos Hip and Hop are playing by hopping over a stone, then landing across so that the stone is the midpoint of the segment traveled during each jump. Picture 1 shows how Hop jumped three times hopping over stones marked 1, 2, and 3 in order. Hip has the same configuration of stones marked 1, 2, and 3 and jumps over them in this order, but starts in a different place as shown in Picture 2. Which of the points A, B, C, D, or E is his landing point?

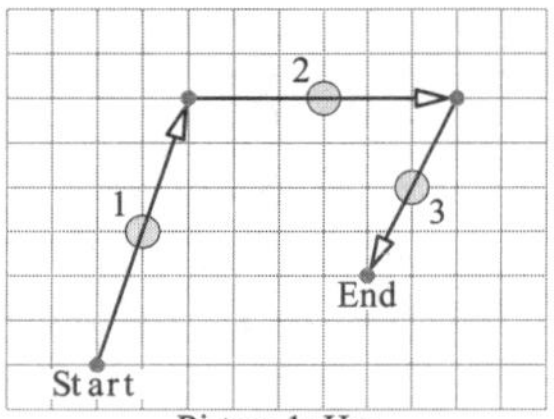

Picture 1: Hop

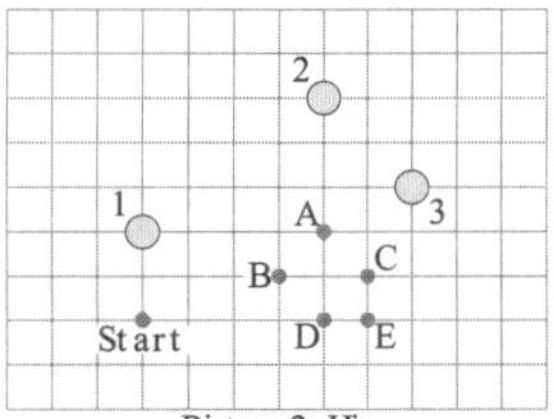

Picture 2: Hip

(A) A (B) B (C) C (D) D (E) E

23. There were twelve children at a birthday party. Each child was either 6, 7, 8, 9, or 10 years old, with at least one child of each age. Four of them were 6 years old. In the group the most common age was 8 years old. What was the average age of the twelve children?

(A) 6 (B) 6.5 (C) 7 (D) 7.5 (E) 8

24. Rectangle $ABCD$ is cut into four smaller rectangles, as show in the figure. The four smaller rectangles have the properties:
(a) the perimeters of three of them are 11, 16, and 19;
(b) the perimeter of the fourth is neither the biggest nor the smallest of the four.
What is the perimeter of the original rectangle $ABCD$?

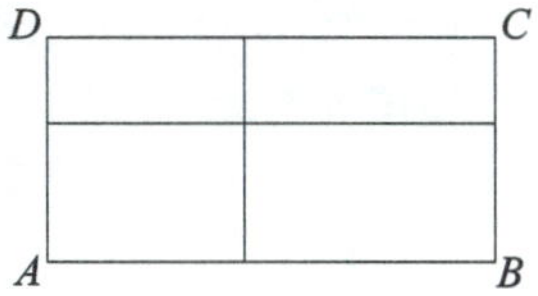

(A) 28 (B) 30 (C) 32 (D) 38 (E) 40

25. Kanga wants to arrange the twelve numbers from 1 to 12 in a circle in such a way that any neighboring numbers always differ by either 1 or 2. Which of the following pairs of numbers have to be neighbors?

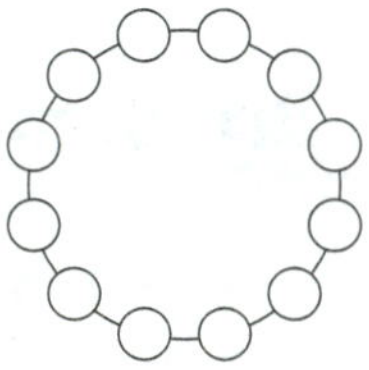

(A) 5 and 6 (B) 10 and 9 (C) 6 and 7 (D) 8 and 10 (E) 4 and 3

26. Peter wants to cut a rectangle of size 6×7 into squares with integer sides. What is the smallest number of squares he can get?

(A) 4 (B) 5 (C) 7 (D) 9 (E) 42

27. Some cells of a square table of size 4×4 were colored red. The number of red cells in each row was indicated at the end of it, and the number of red cells in each column was indicated at the bottom of it. Then the red color was eliminated. Which of the following tables can be the result?

(A) 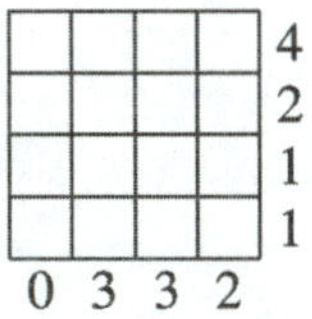

(B) 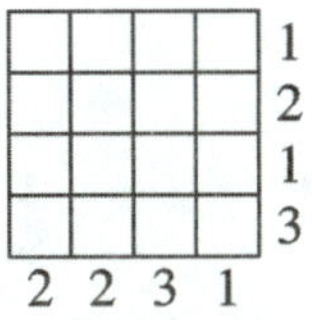

(C) 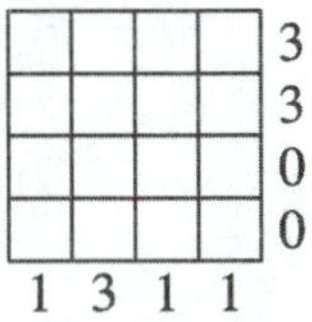

(D) 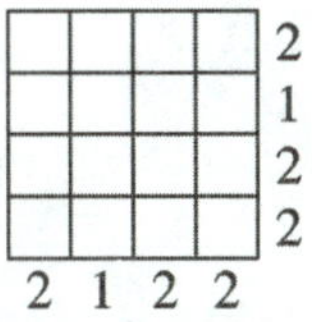

(E) 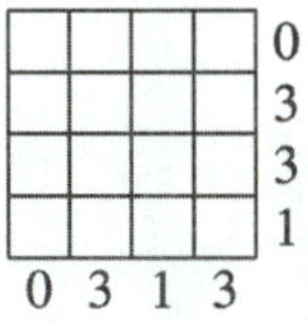

28. A square-shaped piece of paper has an area of 64 cm^2. The square is folded twice as shown in the picture. What is the sum of the areas of the shaded rectangles?

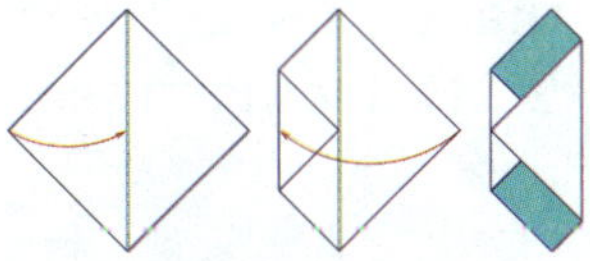

(A) 10 cm^2 (B) 14 cm^2 (C) 15 cm^2 (D) 16 cm^2 (E) 24 cm^2

29. Abid's house number has 3 digits. Removing the first digit of Abid's house number, you obtain Ben's house number. Removing the first digit of Ben's house number, you get Chiara's house number. Adding the house numbers of Abid, Ben, and Chiara gives 912. What is the second digit of Abid's house number?

(A) 3 (B) 4 (C) 5 (D) 6 (E) 0

30. I gave Ann and Bill two consecutive positive integers (for instance 7 to Ann and 6 to Bill). They know their numbers are consecutive, they know their own number, but they do not know the number I gave to the other one. Then I heard the following discussion. Ann said to Bill: "I don't know your number." Bill said to Ann: "I don't know your number." Then Ann said to Bill: "Now I know your number! It is a divisor of 20." What is Ann's number?

(A) 2 (B) 3 (C) 4 (D) 5 (E) 6

Problems from Year 2014

Problems 3 points each

1. Arnold spelled the word KANGAROO with cards showing one letter at a time. Unfortunately, some cards were rotated. By turning the K card back by 90° twice he can correct the letter K, and by turning the first A card once he can correct the first A (see the figures). How many times does he need to rotate by 90° for all of the letters to be correct?

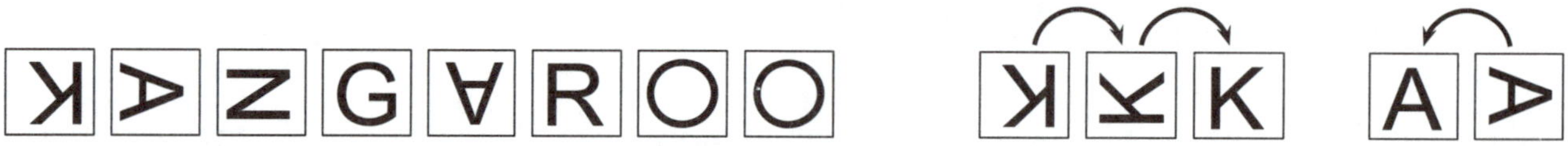

(A) 4 (B) 5 (C) 6 (D) 7 (E) 8

2. A cake weighs 900 g. Paul cuts it in 4 pieces. The biggest piece weighs as much as the 3 other pieces weigh together. What is the weight of the biggest piece?

(A) 250 g (B) 300 g (C) 400 g (D) 450 g (E) 600 g

3. Two large rings, one gray and one white, are linked together. Peter, in front of the rings, sees the rings as in the picture to the right. Paul is behind the rings. What does Paul see?

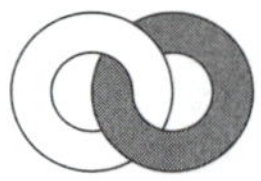

(A) 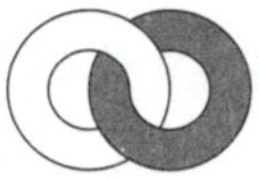 (B)  (C)  (D) 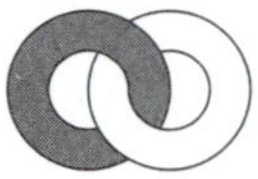 (E) 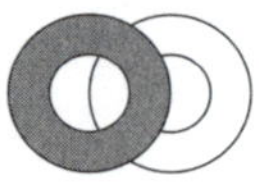

4. In the addition problem to the right, some of the digits have been replaced by stars. What is the sum of the missing digits?

$$\begin{array}{r} 1\star 2 \\ 1\star 3 \\ +\ 1\star 4 \\ \hline 309 \end{array}$$

(A) 0 (B) 1 (C) 2 (D) 3 (E) 10

5. What is the difference between the smallest 5-digit number and the largest 4-digit number?

(A) 1 (B) 10 (C) 1111 (D) 9000 (E) 9900

6. A square with a perimeter of 48 cm (shown on the left of the picture) is cut into 2 pieces. The two pieces are then placed next to each other to make a rectangle (shown on the right of the picture). What is the perimeter of the rectangle?

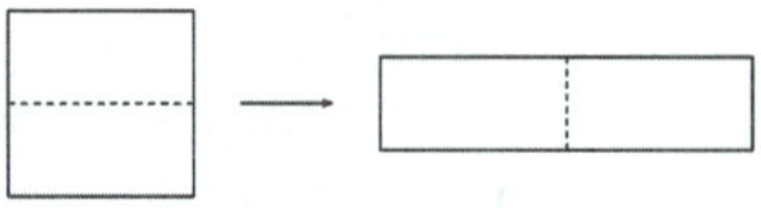

(A) 24 cm (B) 30 cm (C) 48 cm (D) 60 cm (E) 72 cm

7. Katrina has 38 matches. Using all the matches, she builds a triangle and a square. Each side of the triangle consists of 6 matches. How many matches are used for one side of the square?

(A) 4 (B) 5 (C) 6 (D) 7 (E) 8

8. The pearl necklace in the picture contains dark gray pearls and shiny white pearls.

Alex wants to have 5 of the dark gray pearls. She can take pearls from either end of the necklace, and so she has to take some of the white pearls as well. What is the smallest number of white pearls Alex has to take?

(A) 2 (B) 3 (C) 4 (D) 5 (E) 6

9. Harry participated in a broom flight contest which consisted of 5 laps. The times when Harry passed the starting point are shown in the table. Which lap took the shortest time?

	Time
start	09:55
after lap 1	10:26
after lap 2	10:54
after lap 3	11:28
after lap 4	12:03
after lap 5	12:32

(A) the first (B) the second (C) the third (D) the fourth (E) the fifth

10. Ben's digital watch is not working properly. The three horizontal lines in the digit on the far right on the watch do not display. Ben is looking at his watch and the time has just changed from the one shown on the left to the one shown on the right. What time is it now?

(A) 12:40 (B) 12:42 (C) 12:44 (D) 12:47 (E) 12:49

Problems 4 points each

11. Which tile must be added to the picture so that the light gray area is as large as the dark gray area?

?

(A)  (B)  (C)  (D)  (E) It is impossible.

12. Henry and John started walking from the same point. Henry went 1 km north, 2 km west, 4 km south, and finally 1 km west. John went 1 km east, 4 km south, and 4 km west. Which of the following must be the final part of John's walk in order to reach the point where Henry ended his walk?

(A) He has already reached the same point.
(B) 1 km north
(C) 1 km north-west
(D) More than 1 km north-west
(E) 1 km west

13. At the summer camp, 7 children eat ice cream every day, 9 children eat ice cream only every other day, and the rest of the children don't eat ice cream at all. Yesterday, 13 children ate ice cream. How many children will eat ice cream today?

(A) 7 (B) 8 (C) 9 (D) 10 (E) It cannot be determined.

14. Kangaroos A, B, C, D, and E are sitting in that order, clockwise, around a round table. Exactly when the bell rings, each kangaroo except for one exchanges its place with a neighbor. The resulting order, clockwise and starting with kangaroo A, is A, E, B, D, C. Which kangaroo did not move?

(A) A (B) B (C) C (D) D (E) E

15. A square can be formed using four of these five pieces. Which one will not be used?

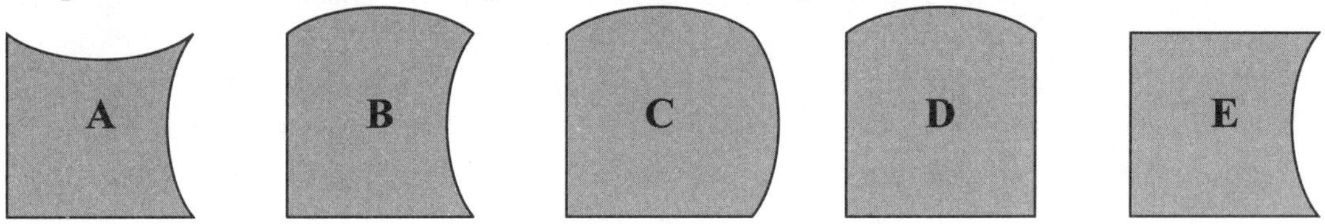

(A) A (B) B (C) C (D) D (E) E

16. A certain natural number has three digits. When we multiply the digits, we get 135. What result do we get if we add the digits?

(A) 14 (B) 15 (C) 16 (D) 17 (E) 18

17. In a restaurant there are 16 tables, each with 3, 4, or 6 chairs. Together, the tables that have 3 or 4 chairs can accommodate 36 people. If the restaurant can accommodate 72 people, how many tables are there with 3 chairs?

(A) 4 (B) 5 (C) 6 (D) 7 (E) 8

18. The points A, B, C, D, E, and F are on a straight line in that order. We know that AF = 35, AC = 12, BD = 11, CE = 12, and DF = 16. What is the distance BE?

(A) 13 (B) 14 (C) 15 (D) 16 (E) 17

19. Marisa set her stones in groups on the desk. After she arranged the stones in groups of 3, she noticed that there were 2 stones left. Then she arranged the stones in groups of 5, and again there were 2 stones left. At least how many more stones does she need so that there won't be any left when she arranges them in groups of 3 and when she arranges them in groups of 5?

(A) 3 (B) 1 (C) 4 (D) 10 (E) 13

20. The faces of a cube are numbered 1, 2, 3, 4, 5, and 6. The faces numbered 1 and 6 have a common edge. The same is true for faces numbered 1 and 5, faces numbered 1 and 2, faces numbered 6 and 5, faces numbered 6 and 4, and faces numbered 6 and 2. Which number is on the face opposite the face with number 4?

(A) 1 (B) 2 (C) 3 (D) 5 (E) It cannot be determined.

Problems 5 points each

21.  The $3 \times 3 \times 3$ cube in the picture is made of 27 small cubes. How many small cubes do you have to take away to see the picture on the right as the result when looking from the right, from above, and from the front? 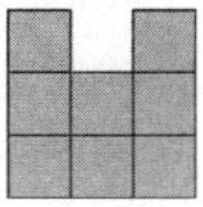

(A) 4 (B) 5 (C) 6 (D) 7 (E) 9

22. There are 5 songs: song A lasts 3 minutes, song B 2 minutes and 30 seconds, song C 2 minutes, song D 1 minute 30 seconds, and song E 4 minutes. These 5 songs are playing in the order A, B, C, D, E in a loop without any breaks. Song C was playing when Andy left home. He returned home exactly one hour later. Which song was playing when Andy got home?

(A) A (B) B (C) C (D) D (E) E

23. Dan entered the numbers 1 to 9 in the cells of a 3×3 table. He began by placing the numbers 1, 2, 3, and 4 as shown in the picture. After he was finished, the sum of the numbers in the cells adjacent to (having a common side with) the cell with the number 5 is equal to 9. What is the sum of the numbers in the cells adjacent to the cell with the number 6?

1		3
2		4

(A) 14 (B) 15 (C) 17 (D) 28 (E) 29

24. Trees grow only on one side of Park Avenue. There are 60 trees in total. Every other tree is a maple, and every third tree is either a linden or a maple. The rest of the trees are birches. How many birches are there?

(A) 10 (B) 15 (C) 20 (D) 24 (E) 30

25. A thin colorful ribbon is glued on a transparent plastic cube (see the picture). Which of the following pictures doesn't show the cube from any perspective?

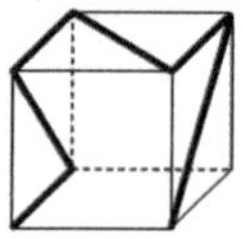

(A)  (B)  (C)  (D)  (E) 

26. The king and his messengers are traveling from the castle to the summer palace at a speed of 5 km/h. Every hour, the king sends a messenger back to the castle, who travels at a speed of 10 km/h. What is the time interval between any two consecutive messengers arriving at the castle?

(A) 30 min (B) 60 min (C) 75 min (D) 90 min (E) 120 min

27. There were 3 one-digit numbers on the blackboard. Ali added them up and got 15. Then he erased one of the numbers and wrote the number 3 in its place. Then Reza multiplied the three numbers on the blackboard and got 36. What are the possibilities for the number that Ali erased?

(A) either 6 or 7 (B) either 7 or 8 (C) only 6 (D) only 7 (E) only 8

28. Peter Rabbit loves cabbages and carrots. In a day, he eats 9 carrots only, 2 cabbages only, or 1 cabbage and 4 carrots. But some days he only eats grass. Over the last 10 days, Peter ate a total of 30 carrots and 9 cabbages. On how many of these 10 days did he eat only grass?

(A) 0 (B) 1 (C) 2 (D) 3 (E) 4

29. In Fabuland, every sunny day is immediately preceded by two consecutive rainy days. Also, five days after any rainy day, it is another rainy day. It is sunny today. For how many days at most can we predict the weather with certainty?

(A) 1 day (B) 2 days (C) 4 days
(D) We cannot predict even one day ahead.
(E) We can predict the weather every day from here on.

30. Granny has 10 grandchildren. Alice is the oldest. One day Granny notices that her grandchildren all have different ages. If the sum of her grandchildren's ages is 180, what is the youngest Alice can be?

(A) 19 (B) 20 (C) 21 (D) 22 (E) 23

Problems from Year 2016

Problems 3 points each

1. Which of the following traffic signs has the largest number of lines of symmetry?

(A)  (B)  (C)  (D)  (E) 

2. Mike cuts pizza into quarters. Then he cuts every quarter into thirds. What part of the whole pizza is one piece?

(A) a third (B) a quarter (C) a seventh (D) an eighth (E) a twelfth

3. A thread with a length of 10 cm is folded into equal parts as shown in the figure. The thread is cut at the two marked places. What are the lengths of the three parts?

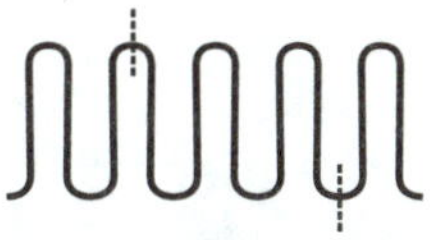

(A) 2 cm, 3 cm, 5 cm (B) 2 cm, 2 cm, 6 cm (C) 1 cm, 4 cm, 5 cm
(D) 1 cm, 3 cm, 6 cm (E) 3 cm, 3 cm, 4 cm

4. On Lisa's refrigerator, 8 strong magnets (the black circles in the picture to the right) are holding up some postcards. What is the largest number of magnets that she can remove so that no postcard falls to the ground?

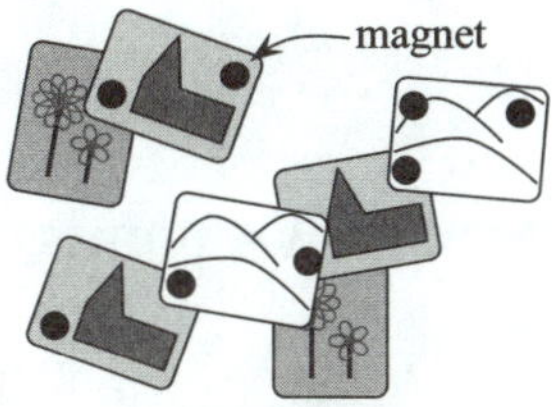

(A) 2 (B) 3 (C) 4 (D) 5 (E) 6

5. Cathy draws a square with a side length of 10 cm. She joins the midpoints of the sides to make a smaller square. What is the area of the smaller square?

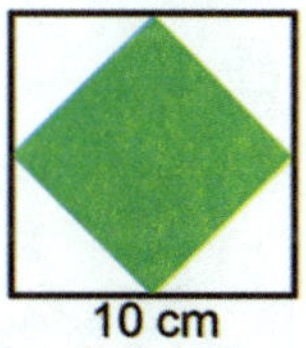

(A) 10 cm^2 (B) 20 cm^2 (C) 25 cm^2 (D) 40 cm^2 (E) 50 cm^2

6. Alice's mother wants to see a knife on the right side of each plate and a fork on the left side. How many interchanges of a knife and a fork does Alice need to make in order to please her mother?

(A) 1 (B) 2 (C) 3 (D) 5 (E) 6

7. A centipede has 25 pairs of shoes. It needs one shoe for each of its 100 feet. How many more shoes does the centipede need to buy?

(A) 15 (B) 20 (C) 35 (D) 50 (E) 75

8. Tom and John are building rectangular boxes using the same number of identical cubes.

Tom's box looks like this: 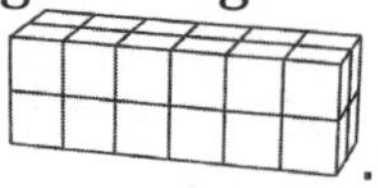. The first level of John's box looks like this: 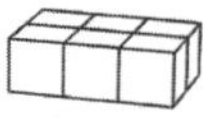.
How many levels will John's box have?

(A) 2 (B) 3 (C) 4 (D) 5 (E) 6

9. In the room shown in the figure to the right, there are four beds with pillows placed as shown by the dark ovals. A girl is sleeping in each of the beds, either on her right side or on her left side. On the left side of the room, Bea and Pia are sleeping with their heads on their pillows and facing each other. On the right side of the room, Mary and Karen are sleeping with their heads on their pillows and with their backs to each other. How many girls are lying on their right sides?

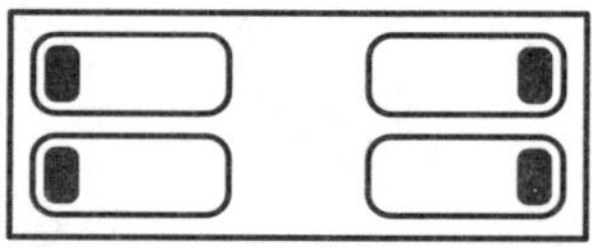

(A) 0 (B) 1 (C) 2 (D) 3 (E) 4

10. The piece of paper shown on the right is folded along the dotted lines to make an open box. The box is put on a table with the top open. Which face is at the bottom of the box?

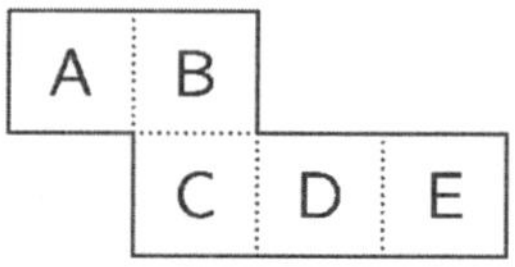

(A) A (B) B (C) C (D) D (E) E

Problems 4 points each

11. Which of the following figures cannot be formed by gluing these two identical squares of paper together?

(A)  (B)  (C)  (D)  (E) 

12. Mary, Ann, and Nata work at a kindergarten. Each day from Monday to Friday exactly two of them come to work. Mary works 3 days per week and Ann works 4 days per week. How many days per week does Nata work?

(A) 1 (B) 2 (C) 3 (D) 4 (E) 5

13. Five squirrels, *A, B, C, D*, and *E*, are sitting on the line. They are going to pick up the 6 nuts, each marked with X. At the same moment each of the squirrels starts running to the nearest nut at the same speed. As soon as a squirrel picks up a nut it starts running to the next closest nut. Which squirrel will get two nuts?

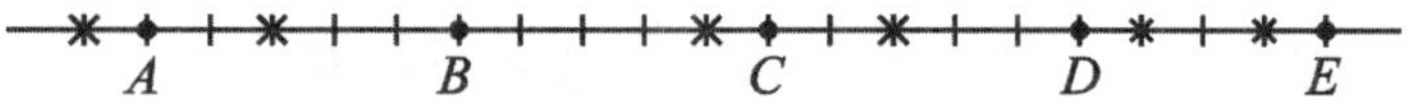

(A) *A* (B) *B* (C) *C* (D) *D* (E) *E*

14. There are 30 students in a class. They sit in pairs so that each boy is sitting with a girl, and exactly half of the girls are sitting with a boy. How many boys are there in the class?

(A) 25 (B) 20 (C) 15 (D) 10 (E) 5

15. The number 2581953764 is written on a strip of paper. John cuts the strip 2 times and gets 3 numbers. Then he adds these 3 numbers. Which is the smallest possible sum he can get?

(A) 2675 (B) 2975 (C) 2978 (D) 4217 (E) 4298

16. Bart is getting his hair cut. When he looks in a mirror at the reflection of the clock behind him, the clock looks like this:

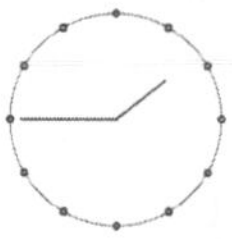

What would he have seen if he had looked in the mirror ten minutes earlier?

(A)  (B) 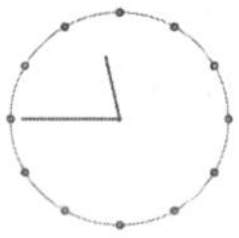 (C)  (D)  (E) 

17. Grandmother bought enough cat food for her four cats to last for 12 days. On her way home she picked up two more cats from the shelter. If she gives each cat the same amount of food every day, how many days will the cat food last?

(A) 8 (B) 7 (C) 6 (D) 5 (E) 4

18. Each letter in BENJAMIN represents one of the digits 1, 2, 3, 4, 5, 6, or 7. Different letters represent different digits. The number BENJAMIN is odd and divisible by 3. Which digit corresponds to N?

(A) 1 (B) 2 (C) 3 (D) 5 (E) 7

19. Tim, Tom, and Jim are triplets, while their brother Carl is 3 years younger. Which of the following numbers could be the sum of the ages of the four brothers?

(A) 53 (B) 54 (C) 56 (D) 59 (E) 60

20. The perimeter of the rectangle $ABCD$ is 30 cm. Three other rectangles are placed so that their centers are at the points A, B, and D (see the figure). The sum of their perimeters is 20 cm. What is the total length of the thick line?

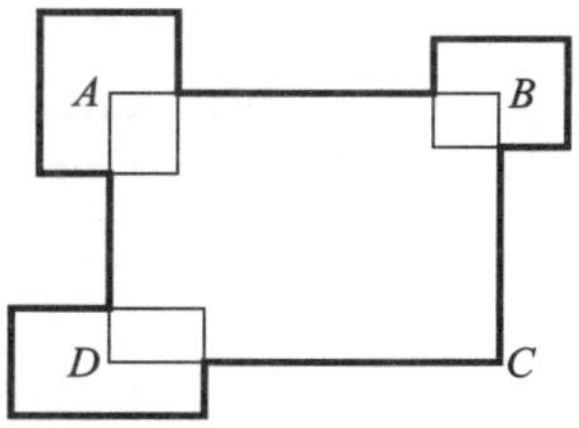

(A) 50 cm (B) 45 cm (C) 40 cm (D) 35 cm
(E) This is impossible to determine.

Problems 5 points each

21. Anna folds a round sheet of paper along the middle line. Then she folds it once more and then one last time.

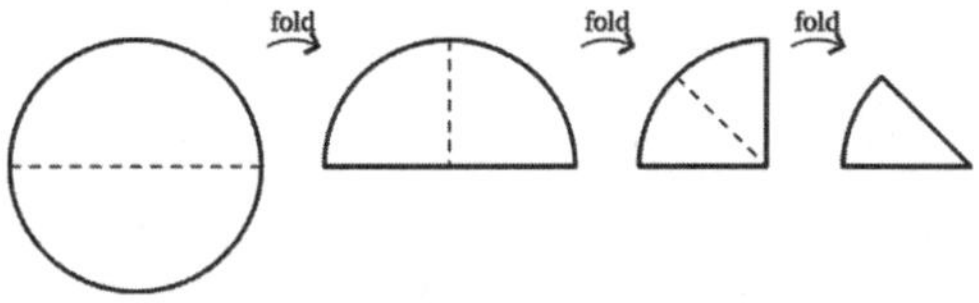

In the end Anna cuts the folded paper along the marked line: 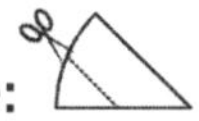. What is the shape of the middle part of the paper when unfolded?

(A)  (B)  (C)  (D)  (E) 

22. Richard writes down all the numbers with the following properties: the first digit is 1, each of the following digits is at least as large as the one before it, and the sum of the digits is 5. How many numbers does he write?

(A) 4 (B) 5 (C) 6 (D) 7 (E) 8

23. What is the greatest number of shapes of the form 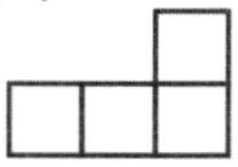 that can be cut out from a 5×5 square?

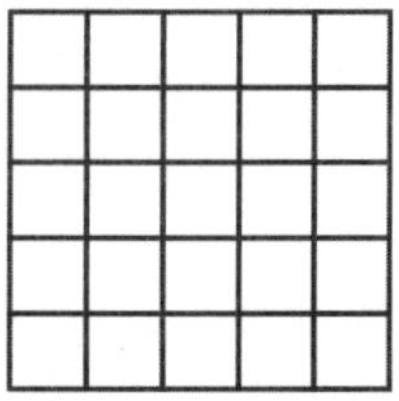

(A) 2 (B) 4 (C) 5 (D) 6 (E) 7

24. Luigi started a small restaurant. His friend Giacomo gave him some square tables and chairs. If Luigi uses all the tables as single tables with 4 chairs each, he will need 6 more chairs. If he uses all the tables as double tables with 6 chairs each, he will have 4 chairs left over. How many tables did Luigi get from Giacomo?

(A) 8 (B) 10 (C) 12 (D) 14 (E) 16

25. Clara wants to construct a big triangle using identical small triangular tiles. She has already put some tiles together as shown in the picture. What is the smallest number of tiles she needs to complete a triangle?

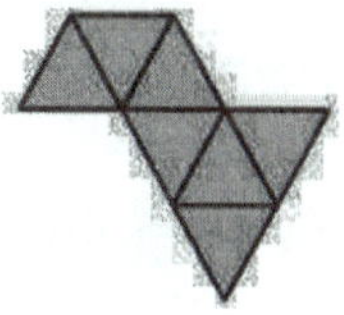

(A) 5 (B) 9 (C) 12 (D) 15 (E) 18

26. A big cube was built from 8 identical small cubes, some black and some white. Five faces of the big cube are:

What does the sixth face of the big cube look like?

(A)  (B)  (C)  (D) 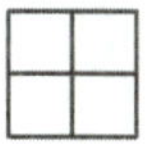 (E) 

27. Kirsten wrote numbers in 5 of the 10 circles as shown in the figure. She wants to write a number in each of the remaining 5 circles such that the sums of the 3 numbers along each side of the pentagon are equal. Which number will she have to write in the circle marked by X?

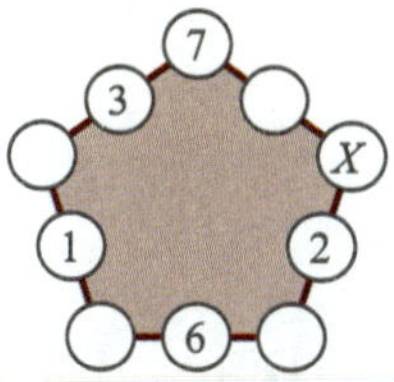

(A) 7 (B) 8 (C) 11 (D) 13 (E) 15

28. The symbols ○, □, and △ represent 3 different digits. If you add the digits of the 3-digit number ○□○, the result is the 2-digit number □△. If you add the digits of the 2-digit number □△, you get the 1-digit number □. What digit does ○ represent?

(A) 4 (B) 5 (C) 6 (D) 8 (E) 9

29. A little kangaroo is playing with his calculator. He starts with the number 12. He multiplies or divides the number by 2 or 3 (if possible) 60 times total. Which of the following results can he not obtain?

(A) 12 (B) 18 (C) 36 (D) 72 (E) 108

30. Two 3-digit numbers are made using 6 different digits. The first digit of the second number is twice the last digit of the first number. What is the smallest possible sum of two such numbers?

(A) 552 (B) 546 (C) 301 (D) 535 (E) 537

Problems from Year 2018

Problems 3 points each

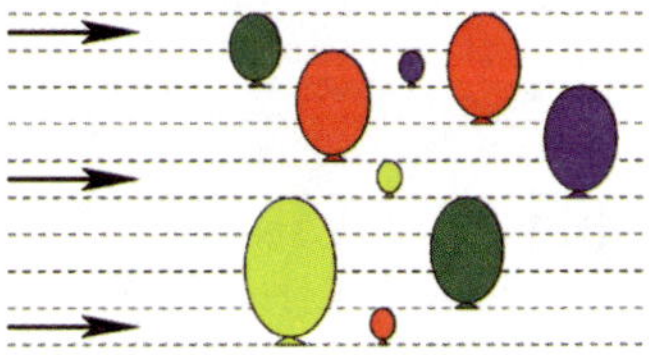

1. The drawing shows 3 flying arrows and 9 balloons that do not move. When an arrow hits a balloon, the balloon pops, and the arrow continues flying farther in the same direction. How many balloons will not be hit by arrows?

(A) 3 (B) 2 (C) 6 (D) 5 (E) 4

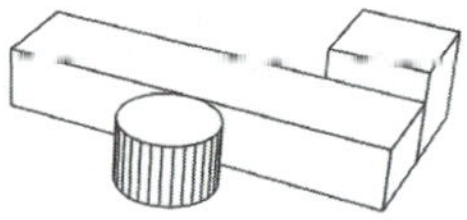

2. There are three objects on the table. What does Peter see if he looks at the table from above?

(A) 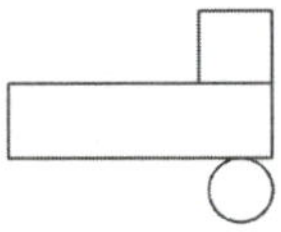 (B) 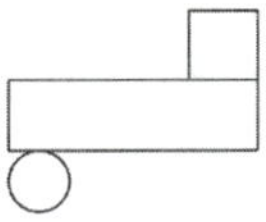 (C) 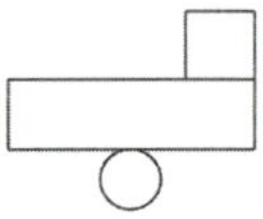 (D) 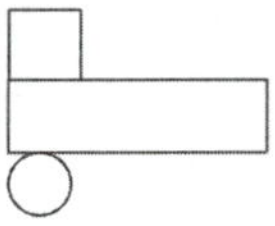 (E) 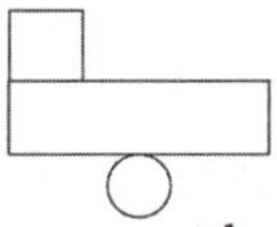

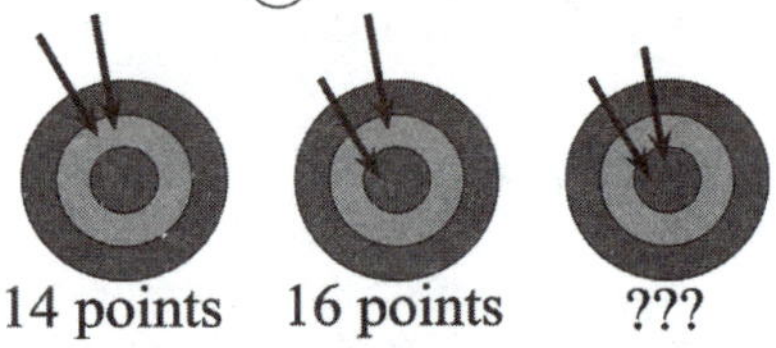

3. Diana first got 14 points with two arrows on the target. The second time she got 16 points. How many points did she get the third time?

(A) 17 (B) 18 (C) 19 (D) 20 (E) 22

4. A garden is divided into identical 1 meter by 1 meter squares. A fast snail and a slow snail move along the perimeter of the garden starting from the corner S but in different directions. The slow snail moves at the speed of 1 meter per hour (1 m/h) and the fast snail at 2 meters per hour (2 m/h). At what point will the two snails meet?

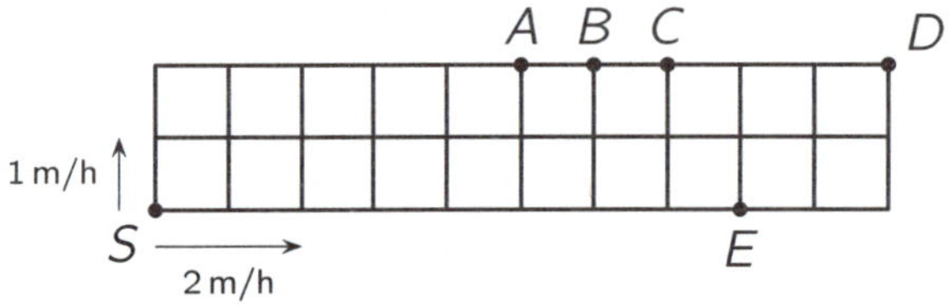

(A) *A* (B) *B* (C) *C* (D) *D* (E) *E*

5. Alice subtracted two 2-digit numbers. Then she painted two cells. What is the sum of the two digits in the painted cells?

(A) 8 (B) 9 (C) 12 (D) 13 (E) 15

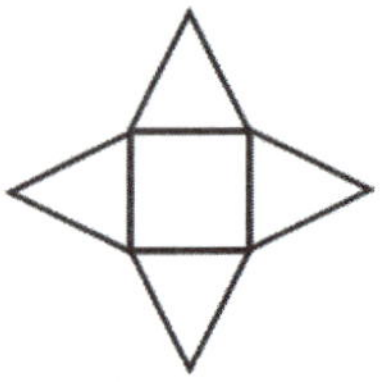

6. A star is made of four equilateral triangles and a square. The perimeter of the square is 36 cm. What is the perimeter of the star?

(A) 144 cm (B) 120 cm (C) 104 cm (D) 90 cm (E) 72 cm

7. The picture shows the calendar of a certain month. Unfortunately, an ink spot covers most of the dates. What day is the 25th of that month?

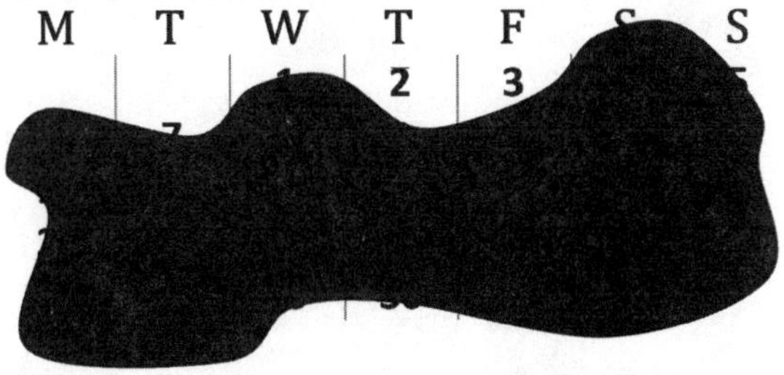

(A) Monday (B) Wednesday (C) Thursday (D) Saturday (E) Sunday

8. How many times do we have to roll a regular six-sided die to be sure that at least one result will be repeated?

(A) 5 (B) 6 (C) 7 (D) 12 (E) 18

9. There are 3 squares in the figure. The side length of the smallest square is 6 cm. What is the side length of the biggest square?

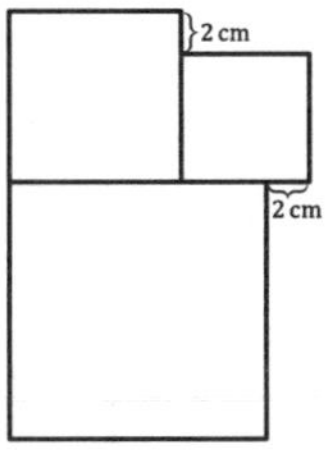

(A) 8 cm (B) 10 cm (C) 12 cm (D) 14 cm (E) 16 cm

10. In the figure on the right, the circles are light bulbs connected to some other light bulbs. Initially, all the light bulbs are off. When you touch a light bulb, this light bulb and all its neighbors are lit. At least how many light bulbs do you have to touch to light all the light bulbs?

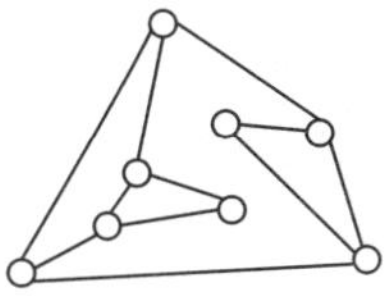

(A) 2 (B) 3 (C) 4 (D) 5 (E) 6

Problems 4 points each

11. In which of the four squares is the ratio of the black area to the white area the largest?

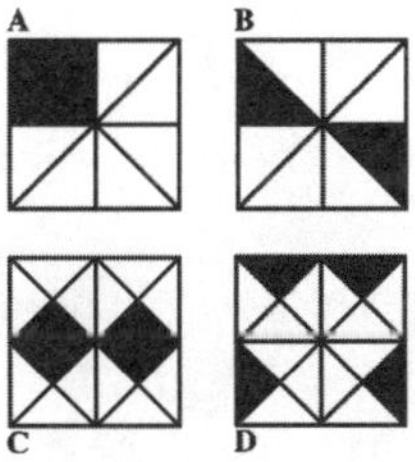

(A) A (B) B (C) C (D) D
(E) They are all the same.

12. Nine cars arrive at an intersection and drive off as indicated by the arrows. Which figure shows these cars after they leave the intersection?

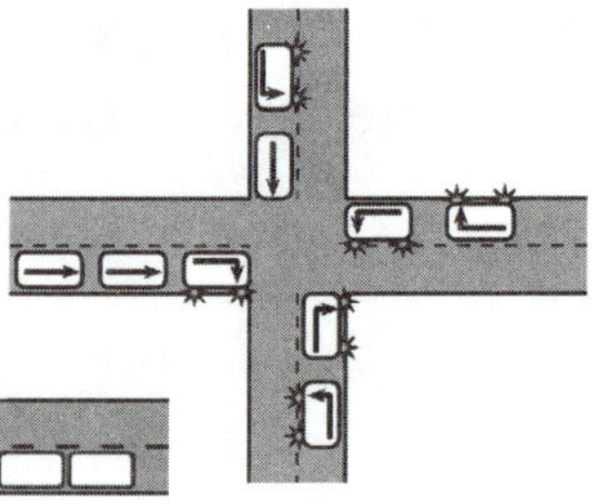

(A) 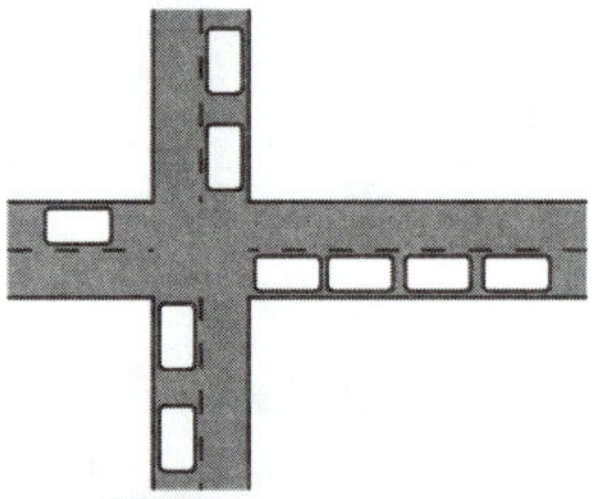 (B) 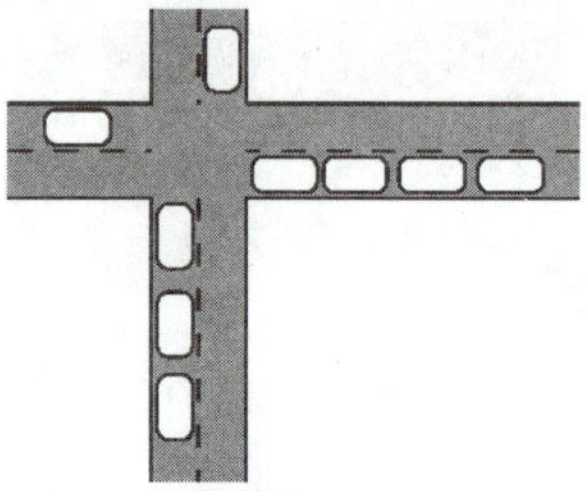 (C) 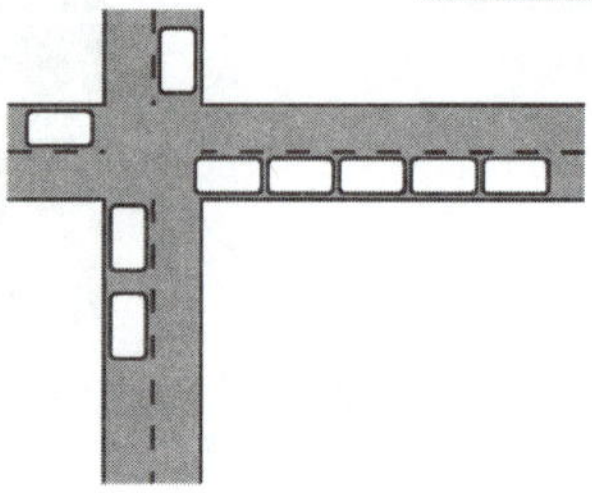

(D) 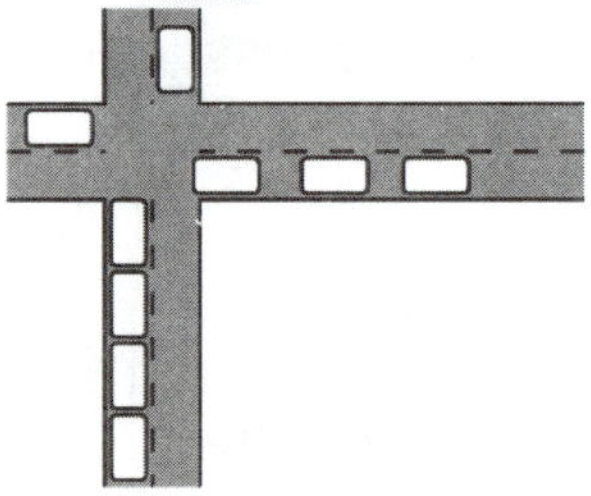 (E) 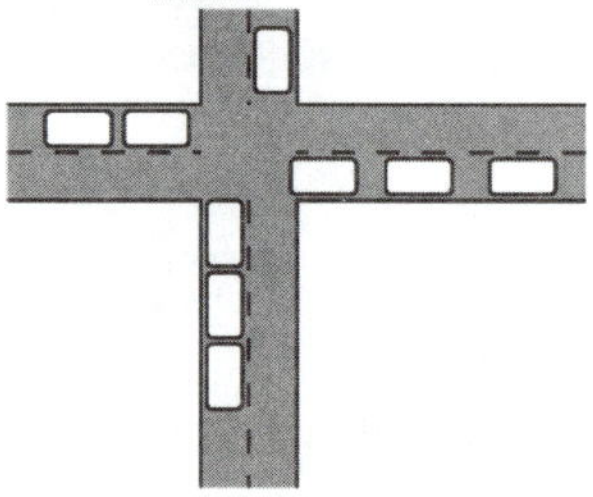

13. Four of the five numbers: 1, 2, 3, 4, and 5, were used so that both calculations following the arrows are correct. The numbers are not shown. Which number is covered by the pink spot with the star?

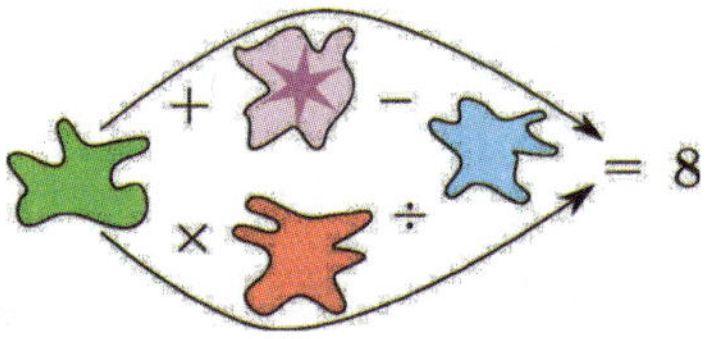

(A) 1 (B) 2 (C) 3 (D) 4 (E) 5

14. A lion is behind one of the three doors. A sentence is written on each door but only one of the three sentences is true. Behind which door is the lion?

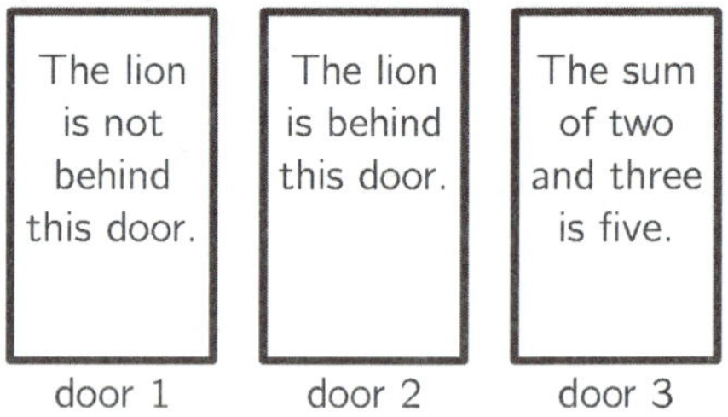

(A) door 1 (B) door 2 (C) door 3
(D) All three doors are possible.
(E) Both door 1 and door 2 are possible.

15. Two girls (Eva and Olga) and three boys (Adam, Isaac, and Urban) are playing with a ball. When a girl has the ball, she throws it to the other girl or to a boy. When a boy has the ball, he throws it to another boy but never to the boy from whom he just received it. Eva starts by throwing the ball to Adam. Who will do the fifth throw?

(A) Adam (B) Eva (C) Isaac (D) Olga (E) Urban

16. Emily wants to enter a number into each cell of the triangular table. The sum of the numbers in any two cells with a common edge must be the same. She has already entered two numbers. What is the sum of all the numbers in the table?

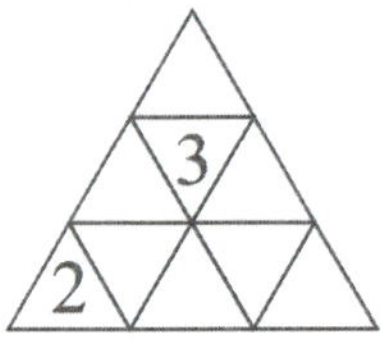

(A) 18 (B) 20 (C) 21 (D) 22
(E) It is impossible to determine.

17. On Monday, Alexandra emails a picture to 5 friends. For several days everybody who receives the picture emails it the next day to two friends who haven't received the picture yet. On which day does the number of people who have received the picture become greater than 100?

(A) Wednesday (B) Thursday (C) Friday (D) Saturday (E) Sunday

18. The faces of a cube are painted black, white, or gray, so that opposite faces are of different color. Which of the following is not a possible net of this cube?

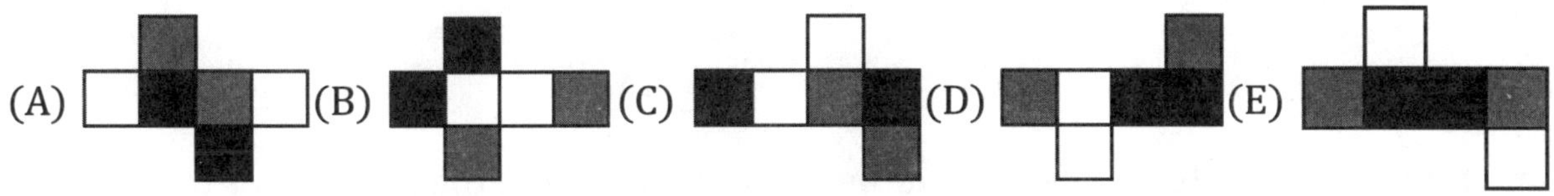

19. John does a calculation using four digits represented in the addition by A, B, C, and D. Which digit is represented by B?

$$\begin{array}{r} ABC \\ +\,CBA \\ \hline DDDD \end{array}$$

(A) 0 (B) 2 (C) 4 (D) 5 (E) 6

20. Four ladybugs are sitting on different cells of a 4×4 grid. One of them is sleeping and does not move. Each time you whistle, the other 3 ladybugs move to a free neighboring cell. They can move up, down, right, or left, but they are not allowed to go back to the cell they just came from.

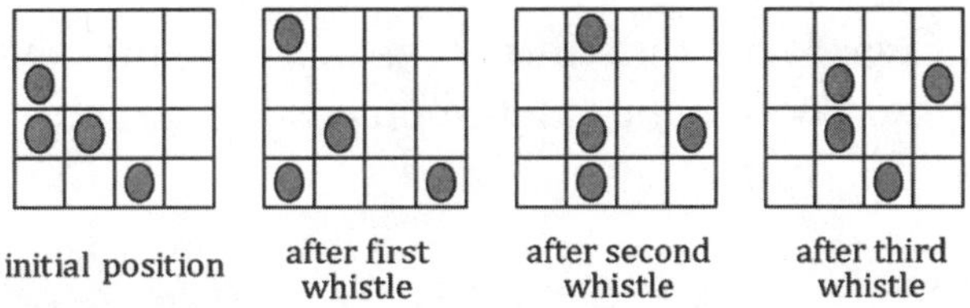

Which of the following images might show the result after the fourth whistle?

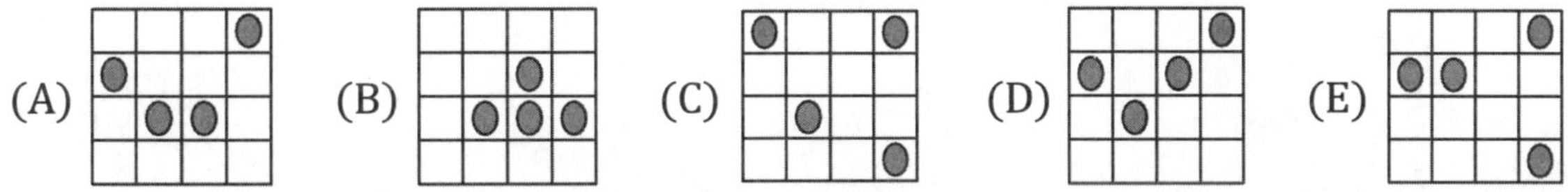

Problems 5 points each

21. From the following numbers: 3, 5, 2, 6, 1, 4, and 7, Masha chose three numbers with a sum of 8. From the same list Dasha chose three numbers with a sum of 7. How many of the numbers chosen by both girls are the same?

(A) none (B) 1 (C) 2 (D) 3 (E) It is impossible to determine.

22. Five balls, A, B, C, D, and E, weigh 30 g, 50 g, 50 g, 50 g, and 80 g each, not necessarily in that order. Which ball weighs 30 g?

(A) A (B) B (C) C (D) D (E) E

23. A, B, and C represent three different digits. The greatest possible 6-digit number that uses digit A three times, digit B two times, and digit C just once will never be written in the form

(A) AAABBC (B) CAAABB (C) BBAAAC (D) AAABCB (E) AAACBB

24. The sum of the ages of Kate and her mother is 36, and the sum of the ages of her mother and her granny is 81. How old was her granny when Kate was born?

(A) 28 (B) 38 (C) 45 (D) 53 (E) 56

25. Nick wants to arrange the numbers 2, 3, 4, 5, 6, 7, 8, 9, and 10 into several groups so that the sum of the numbers in each group is the same. What is the largest number of groups he can make?

(A) 2 (B) 3 (C) 4 (D) 6 (E) other answer

26. Peter cut a long board that was 8 cm wide into 9 pieces. One piece was a square, and the rest were rectangles. Then he put all the pieces together as shown in the picture. How long was the board?

(A) 150 cm (B) 168 cm (C) 196 cm (D) 200 cm (E) 232 cm

27. Write 0 or 1 in each cell of the 5×5 table so that each 2×2 square of the 5×5 table contains exactly 3 equal numbers. What is the largest possible sum of all the numbers in the table?

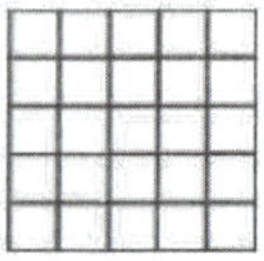

(A) 22 (B) 21 (C) 20 (D) 19 (E) 18

28. 14 people are seated at a round table. Each person is either a liar or tells the truth. Each one of them says, "Both my neighbors are liars." What is the maximum number of liars at the table?

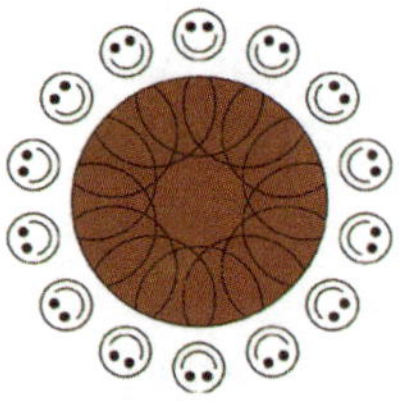

(A) 7 (B) 8 (C) 9 (D) 10 (E) 14

29. There are eight domino tiles on the table (picture 1). One half of one tile is covered. The 8 tiles can be arranged into a 4×4 square (picture 2), so that the number of dots in each row and column is the same. How many dots are on the covered part?

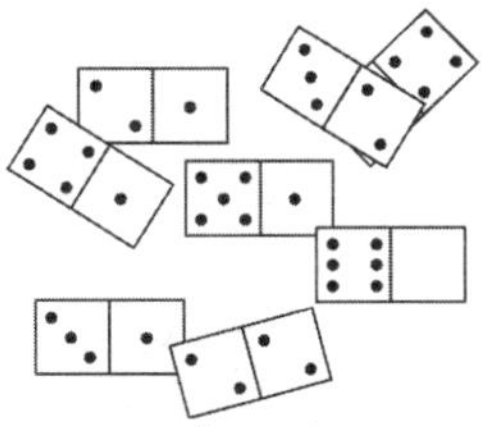
picture 1

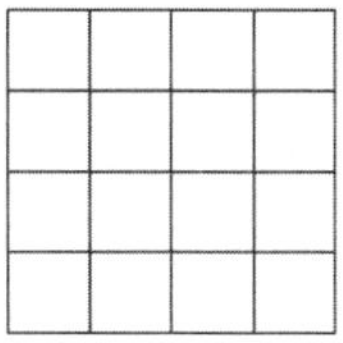
picture 2

(A) 1 (B) 2 (C) 3 (D) 4 (E) 5

30. Write the numbers 3, 4, 5, 6, 7, 8, and 9 in the seven circles to obtain equal sums along each of the three straight lines. What is the sum of all the possible numbers that can be placed in the circle marked with the question mark?

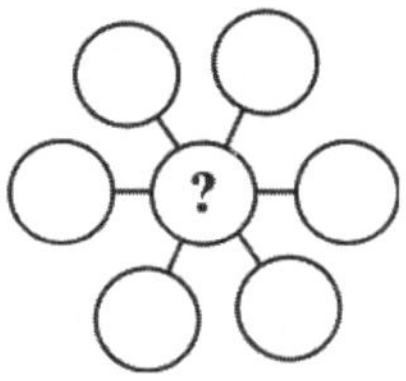

(A) 3 (B) 6 (C) 9 (D) 12 (E) 18

Problems from Year 2020

Problems 3 points each

1. Which piece completes the pattern?

(A)  (B)  (C)  (D)  (E) 

2. As Amira is walking from Atown to Betown she passes the five signposts shown. One of them is incorrect. Which one?

(A) 

(B) 

(C) 

(D) 

(E) 

3. A large square is made up of smaller white and gray squares. What does the large square look like if the colors of the white and gray squares are switched?

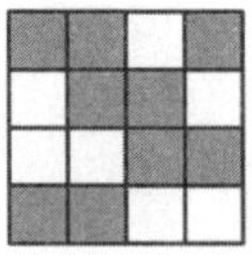

(A) 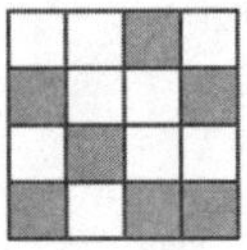 (B) 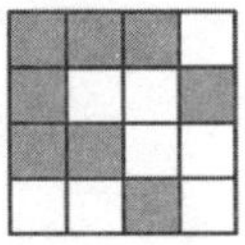 (C) 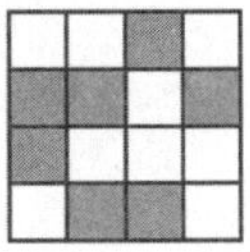 (D) 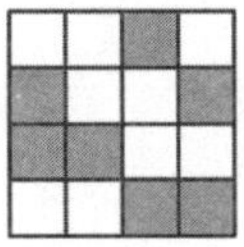 (E) 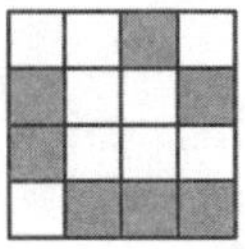

4. Mikas wants to bake 24 muffins for his birthday party. To bake six muffins, he needs two eggs. Eggs are sold in boxes of six. How many boxes does Mikas need to buy?

(A) 1 (B) 2 (C) 3 (D) 4 (E) 5

5. Flora reflects the letter F over the two lines shown:  . What will the reflections look like?

(A)  (B)  (C) 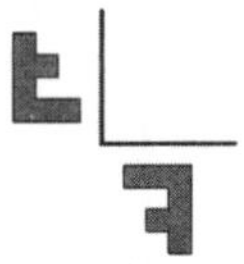 (D) 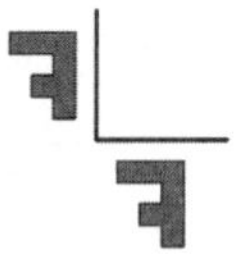 (E) 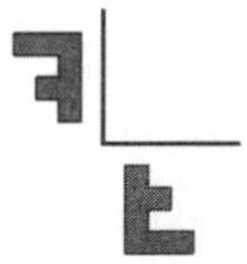

6. Kim has several chains of length 5 and of length 7. By joining chains one after the other, Kim can create different lengths. Which of these lengths is impossible to make?

(A) 10 (B) 12 (C) 13 (D) 14 (E) 15

7. Maria has 10 sheets of paper. She cuts some of the sheets into five parts each. After that, Maria has 22 pieces in total. How many sheets did she cut?

(A) 3 (B) 2 (C) 6 (D) 7 (E) 8

8. Cindy colors each region of the pattern shown to the right red, blue, or yellow. She colors regions that touch each other different colors. She colors the outer region blue. How many regions of the completed pattern are colored blue?

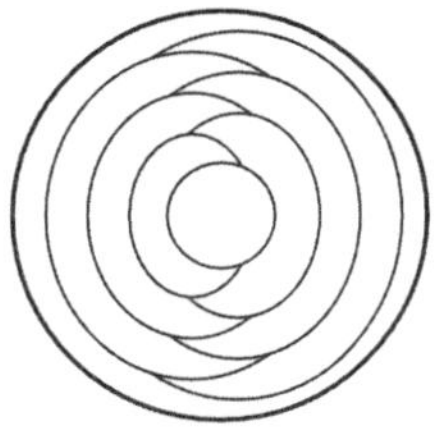

(A) 2 (B) 3 (C) 4 (D) 5 (E) 6

9. Four baskets contain 1, 4, 6, and 9 apples respectively. At least how many apples should be moved between the baskets to have the same number of apples in each basket?

(A) 3 (B) 4 (C) 5 (D) 6 (E) 7

10. A dog and a cat are walking in the park along the path marked by the thick black line. The dog starts from P at the same time as the cat starts from Q. The dog walks three times as fast as the cat. At which point do they meet?

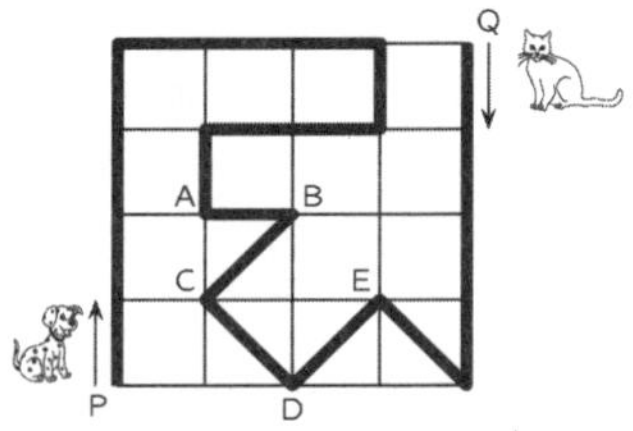

(A) at A (B) at B (C) at C (D) at D (E) at E

Problems 4 points each

11. The numbers from 1 to 10 have to be placed in the small circles, one in each circle. The numbers in any two neighboring circles must have the same sum as the numbers in the two diametrically opposite circles. Some of the numbers are already placed. What number should be placed in the circle with the question mark?

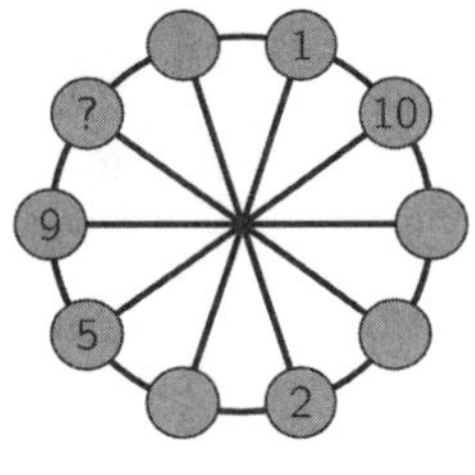

(A) 3 (B) 4 (C) 6 (D) 7 (E) 8

12. When Elise the bat leaves her cave, a digital clock shows 20:20. When she returns and is hanging upside down, she sees 20:20 on the clock again. How long has she been away from her cave?

(A) 3 hours and 28 minutes (B) 3 hours and 40 minutes (C) 3 hours and 42 minutes
(D) 4 hours and 18 minutes (E) 5 hours and 42 minutes

13. An elf and a troll meet. The troll always lies, while the elf always tells the truth. They both say exactly one of the following sentences. Which one?

(A) I am telling the truth. (B) You are telling the truth. (C) We both are telling the truth.
(D) I always lie. (E) One and only one of us is telling the truth.

14. Mary has exactly 10 white cubes, 9 light gray cubes, and 8 dark gray cubes, all of the same size. She glues all these cubes together to build a big cube. One of the cubes below is the one she builds. Which one?

(A)  (B)  (C)  (D)  (E) 

15. The diagrams show five paths from X to Y marked with a thick line. Which path is the shortest?

(A) 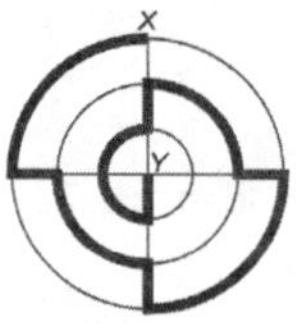

(B) 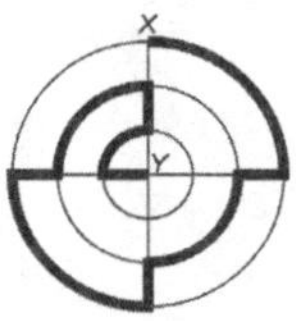

(C) 

(D) 

(E) 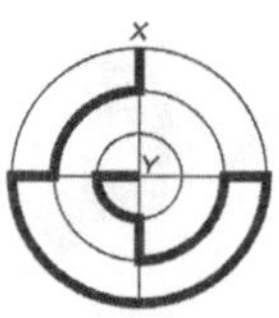

16. A father kangaroo lives with his three children. They decide on all matters by taking a vote, and each member of the family gets as many votes as his or her age. The father is 36 years old and the children are 13, 6, and 4 years old, so right now the father always wins. How many years will it take for the children to have majority of votes if they all vote the same way?

(A) 5 (B) 6 (C) 7 (D) 13 (E) 14

17. Giorgio has two identical pieces of wire of this shape: 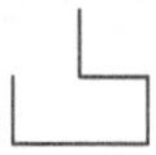. Which of the following shapes cannot be made by putting together these two pieces?

(A)  (B) 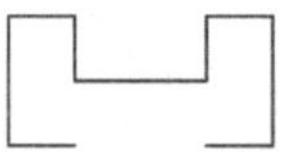 (C) 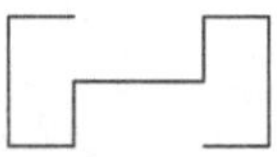 (D) 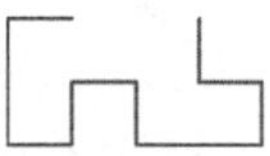 (E) 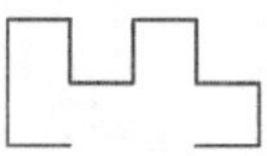

18. Amy glues the six stickers shown onto the faces of a cube: .
The pictures show the cube in two different positions.
Which sticker is on the face opposite the face with the mouse?

(A)  (B)  (C)  (D)  (E) 

19. The diagram shows the friendships of the six girls: Ann, Beatrice, Chloe, Diana, Elizabeth, and Fiona. Each number represents one of the girls and each line joining two numbers represents a friendship between those two girls. Chloe, Diana, and Fiona each have four friends. Beatrice is friends with only Chloe and Diana. Which number represents Fiona?

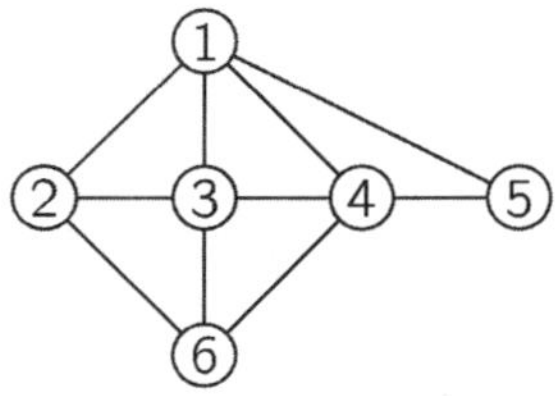

(A) 2 (B) 3 (C) 4 (D) 5 (E) 6

20. Mary put the same amount of liquid in three rectangular vessels. Viewed from the front, they seem to have the same size, but the liquid has risen to different levels in the three vessels. Which of the following images represents the three vessels when viewed from above?

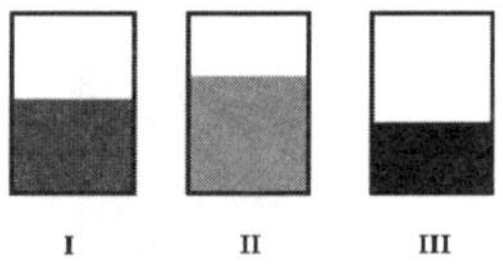

(A) 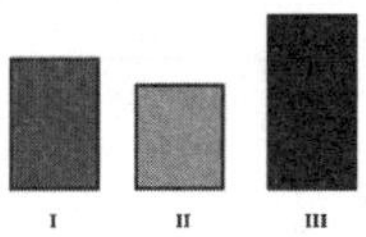 (B) 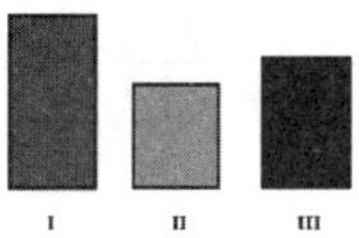 (C) 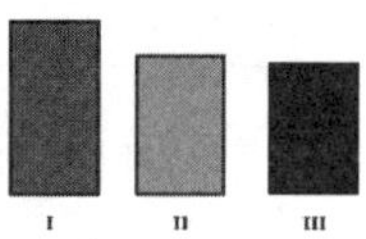 (D) 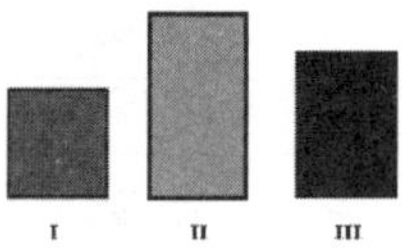 (E) 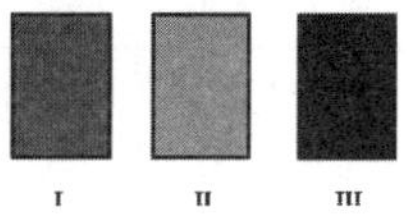

Problems 5 points each

21. What does the object in the picture look like when viewed from above?

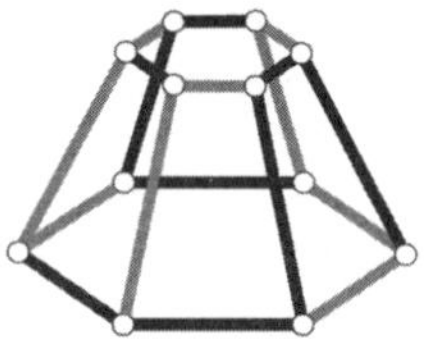

(A) 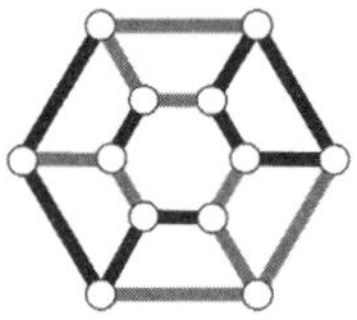 (B) 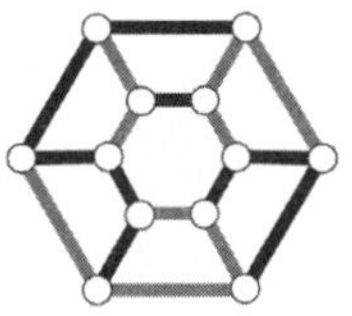 (C) 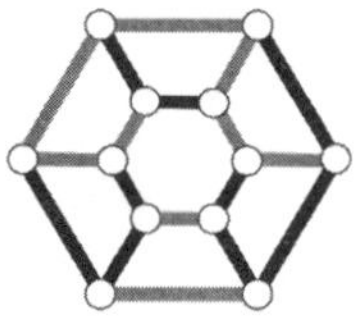 (D) 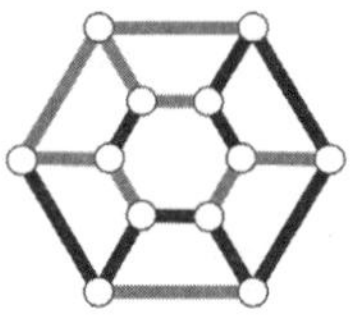 (E) 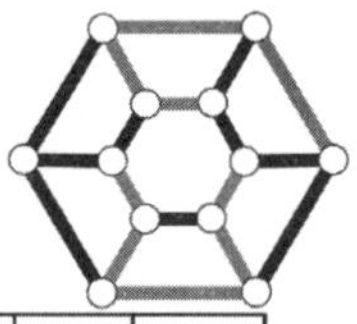

22. Three small squares are drawn inside a larger square as shown. What is the length of the line marked with the question mark?

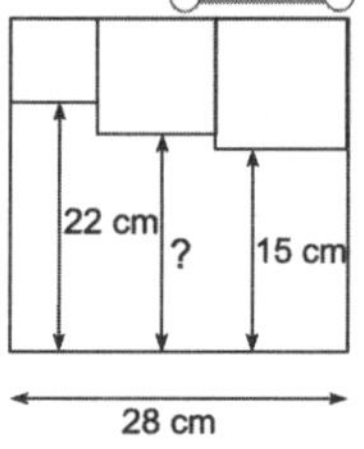

(A) 17 cm (B) 17.5 cm (C) 18 cm (D) 18.5 cm (E) 19 cm

23. Nine tokens are black on one side and white on the other. Initially, four tokens have the black side up: ●●●●○○○○○. In each turn you have to flip three tokens. What is the least number of turns you need to have all tokens showing the same color?

(A) 1 (B) 2 (C) 3 (D) 4 (E) 5

24. Which of the following options will definitely balance the third scale?

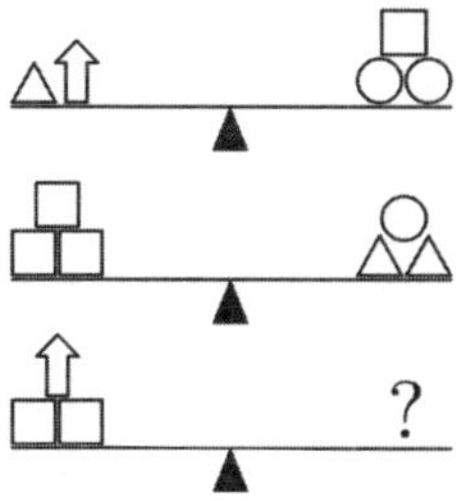

(A) △△△△□ (B) △△△○ (C) △○○○

(D) △□□□□ (E) ○○□

25. Ten people each order one scoop of ice cream. They order 4 scoops of vanilla, 3 scoops of chocolate, 2 scoops of lemon, and 1 scoop of mango. They top the ice cream scoops with 4 umbrellas, 3 cherries, 2 wafers, and 1 chocolate chip. They use one topping on each scoop in such a way that no two ice cream scoops are alike. Which of the following combinations is NOT possible?

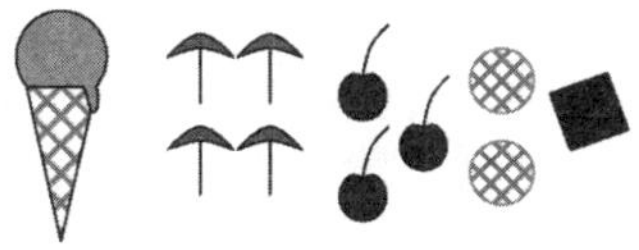

(A) chocolate with a cherry (B) mango with an umbrella (C) vanilla with an umbrella
(D) lemon with a wafer (E) vanilla with a chocolate chip

26. We call a 3-digit number *nice* if its middle digit is greater than the sum of its first and last digits. What is the largest possible number of consecutive *nice* 3-digit numbers?

(A) 5 (B) 6 (C) 7 (D) 8 (E) 9

27. Magnus has to play 15 games in a chess tournament. At a certain point during the tournament, he has won half of the games he has played, he has lost one third of the games he has played, and two have ended in a draw. How many games does Magnus still have to play?

(A) 2 (B) 3 (C) 4 (D) 5 (E) 6

28. Vadim has a square piece of paper divided into nine cells. He folds the paper as shown, overlapping horizontally and then vertically so that the gray square ends up on top. Vadim wants to write the numbers from 1 to 9 into the cells so that once the paper is folded the numbers would be in increasing order with number 1 on the top layer. What numbers should he write instead of a, b, and c?

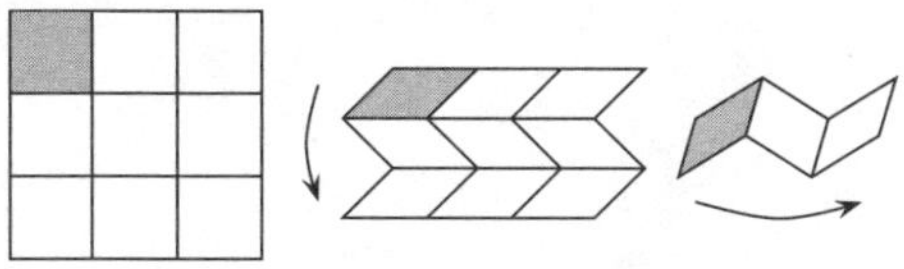

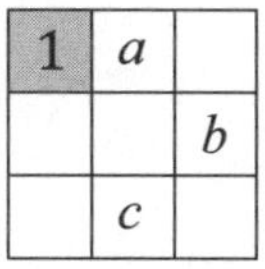

(A) $a = 6,\ b = 4,\ c = 8$ (B) $a = 4,\ b = 6,\ c = 8$ (C) $a = 5,\ b = 7,\ c = 9$
(D) $a = 4,\ b = 5,\ c = 7$ (E) $a = 6,\ b = 4,\ c = 7$

29. Don builds a pyramid using balls. The square base consists of 3×3 balls. The middle layer has 2×2 balls, and there is one ball at the top. Any two balls that touch each other are glued at their contact point. How many glued contact points are there?

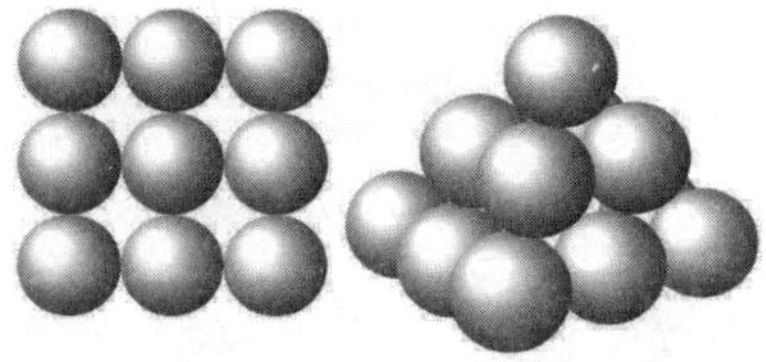

(A) 20 (B) 24 (C) 28 (D) 32 (E) 36

30. The figure shows a map of some islands and how they are connected by bridges. A postman has to visit each island exactly once. He starts on the island marked "start" and would like to finish on the island marked "finish." He has just reached the black island at the center of the map. In which direction should he move to be able to complete his route?

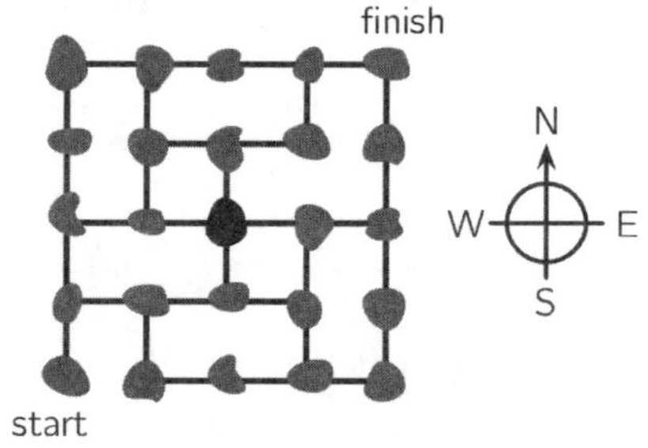

(A) north (B) east (C) south (D) west
(E) There is no such path as the postman wishes to follow.

Part II: Solutions

Solutions for Year 1998

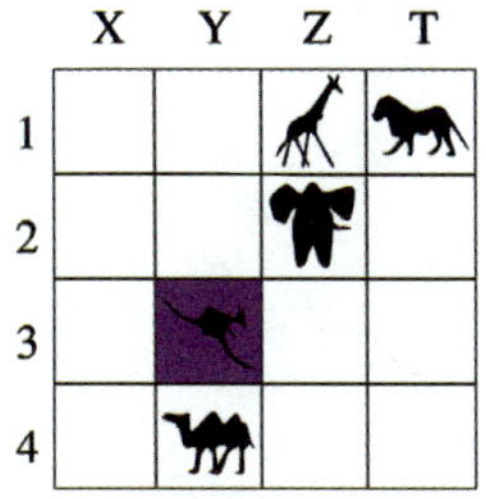

1. (B) 3Y

The kangaroo is in the 3rd row and the Y column, so its coordinates are 3Y.

2. (A) 11 minutes

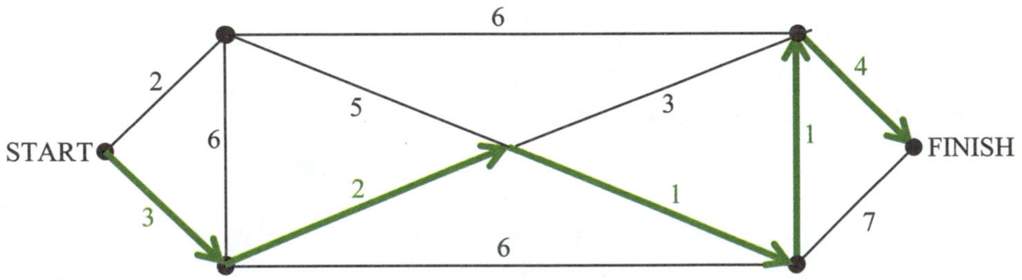

Try different paths. The quickest path is marked by green arrows and the time along this path is $3 + 2 + 1 + 1 + 4$, which is 11 minutes.

3. (B) 1 and 5

The puzzle pieces are formed by identical squares with identical rounded tabs (shown in blue) and cut-outs. Each cut-out removes the area equivalent to one tab.

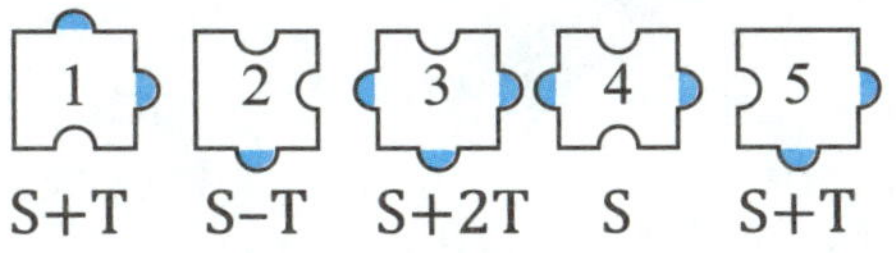

For each puzzle piece, the area is shown underneath. S is the area of one square and T is the area of one tab. Only pieces 1 and 5 have the same area. Notice that piece 5 is simply piece 1 rotated 90° clockwise.

4. (C) 361

$16^2 = 256$, $17^2 = 289$, $18^2 = 324$, $19^2 = 361$, and $20^2 = 400$ are consecutive perfect squares, so 361 is the smallest natural number greater than 360 that is a square of a natural number.

5. (A) 4 h 40 min

One week on Mars is 7×40 min $= 280$ min longer than one week on Earth. $280 = 4 \times 60 + 40$, so 280 min = 4 h 40 min. One week on Mars is 4 h 40 min longer than one week on Earth.

6. (E) 6

The rectangles are identified by their heights. There are 6 of them.

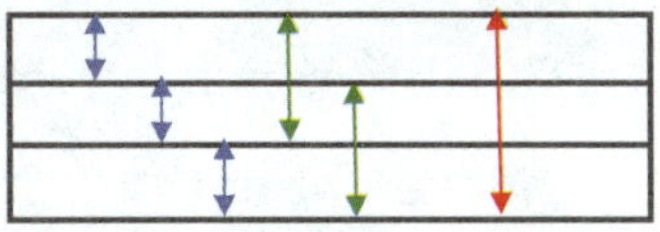

7. (C) 180

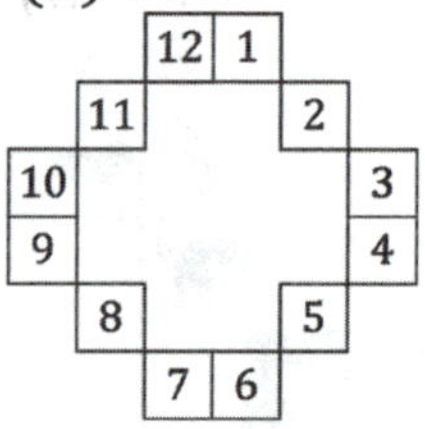

The number of hourly strikes for half a day is shown to the left. In each row the sum of numbers is 13, so the number of all hourly strikes is $6 \times 13 = 78$ for half a day. For one whole day the number of hourly strikes is $2 \times 78 = 156$ and there are also 24 strikes at the half-hour marks, so $156 + 24 = 180$ strikes can be heard in one 24-hour period.

8. (D) 26

Following spring 1998, the Summer Olympics will take place in the years 2000, 2004, 2008, 2012, 2016, 2020, 2024, 2028, 2032, 2036, 2040, 2044, and 2048. The Winter Olympics will take place in the years 2002, 2006, 2010, 2014, 2018, 2022, 2026, 2030, 2034, 2038, 2042, 2046, and 2050. Counting both the summer and winter competitions, the Olympics will take place 26 times between spring 1998 and March 20, 2051.

9. (D) 6

There are 3 ways to keep one pocket empty and 3 ways to keep two pockets empty.

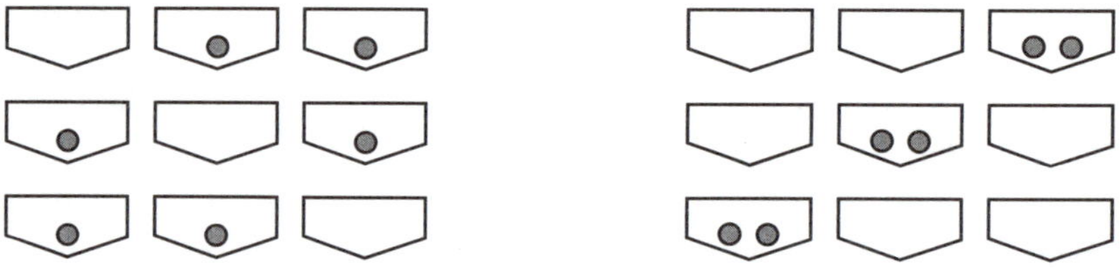

10. (C) ЯUGИAꓘ

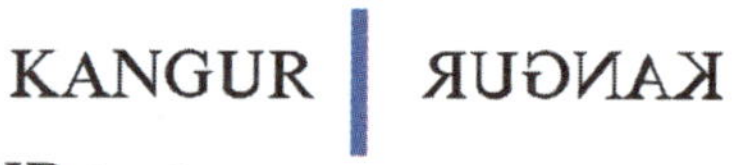

The mirror image of KANGUR is the same as a symmetric image of KANGUR with respect to a line perpendicular to the line of KANGUR and it is shown to the right. Symmetry with respect to a line parallel to the line of KANGUR gives us (A) KVИGՈR. Rotation of KANGUR by 180° gives us (E) ꓤՈ⅁NVꓘ. KANGUR written from right to left is (B) RUGNAK and its mirror image is (D) ꓘAИGUЯ.

11. (B) 7

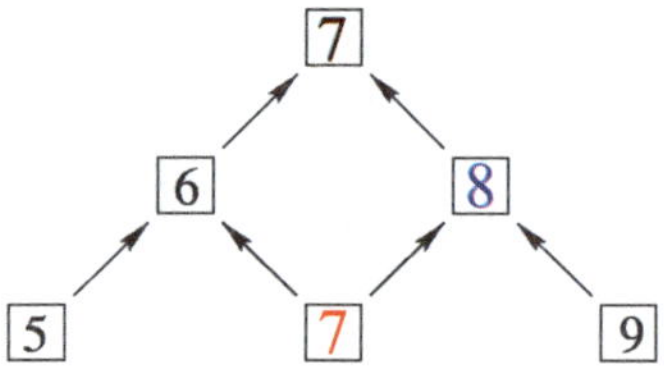

Place 7 (shown in red) in the empty space in the bottom row since $\frac{5+7}{2} = (5 + 7) \div 2 = 12 \div 2 = 6$. Place 8 above 7 and 9 because $\frac{7+9}{2} = 16 \div 2 = 8$. Then, place 7 above 6 and 8 as $\frac{6+8}{2} = 14 \div 2 = 7$. Thus, the number at the top of the pyramid is 7.

12. (B) 4 kg

The difference between weights of 1 watermelon and $\frac{4}{5}$ of the same watermelon is $\frac{1}{5}$ of the weight of the watermelon and that difference is $\frac{4}{5}$ kg. Hence, $\frac{1}{5}$ of the watermelon weighs $\frac{4}{5}$ kg, so one whole watermelon weighs $5 \times \frac{4}{5}$ kg $= 4$ kg.

13. (D) 2
17 is neither a multiple of 4 nor a multiple of 3, so there is at least one **pair** of a chair and a stool. This pair accounts for 7 legs. $17 - 7 = 10$ and again 10 is neither a multiple of 4 nor a multiple of 3, so there is another **pair** of a chair and a stool. These two pairs account for $2 \times 7 = 14$ legs. The other $17 - 14 = 3$ legs belong to one stool, so there are 3 stools and 2 chairs in the room. $3 \times 3 + 2 \times 4 = 17$ is the number of all legs.

14. (C) 110
$\square + \triangle + \bigcirc + \bigcirc = (\square + \bigcirc) + (\triangle + \bigcirc) = 30 + 80 = 110.$
We do not need the middle equation, $\square + \triangle + \triangle = 160$, for the calculation. Also, we don't actually need to figure out what number each shape represents.

15. (D) 9
The difference $(100x + 10y + z) - (100z + 10y + x)$ simplifies to $99x - 99z$ or $99 \times (x - z)$, so the difference is always a multiple of 99, 33, 11, 9, 3, and 1. Only 9 is among the five given options, so 9 is the answer.

16. (C) 48
44 weeks + 44 days + 44 hours = 44 weeks + 7 weeks + 2 days + 1 day + 20 hours, so
44 weeks + 44 days + 44 hours = 51 weeks + 3 days + 20 hours, which is between 51½ and 52 weeks.
One month is between 4 weeks (28 days) and 4½ weeks (31½ days), so 44 months are between $44 \times 4 = 176$ and $44 \times 4½ = 176 + 22 = 198$ weeks. Therefore, all of Mr. Kowalski's months, weeks, days, and hours are between $51½ + 176 = 227½$ and $52 + 198 = 250$ weeks.
1 year is less than 52½ weeks (367½ days), so 4 years is less than $4 \times 52½ = 210$ weeks. $210 < 227½$, so 4 years is less than all of Mr. Kowalski's months, weeks, days, and hours.
1 year is more than 52 weeks (364 days), so 5 years is more than $5 \times 52 = 260$ weeks.
$260 > 250$, so 5 years is more than all of Mr. Kowalski's months, weeks, days, and hours.
In summary, Mr. Kowalski's has lived between $4 + 44 = 48$ years and $5 + 44 = 49$ years.
Therefore, Mr. Kowalski is 48 years old.

17. (D) 8
Let A_1, B_1 be the first couple. Similarly, A_2, B_2 and A_3, B_3 be the 2nd and 3rd couple, respectively. Then the 3-person groups where there will not be a married couple are: $\{A_1,A_2,A_3\}$, $\{A_1,A_2,B_3\}$, $\{A_1,B_2,A_3\}$, $\{A_1,B_2,B_3\}$, $\{B_1,A_2,A_3\}$, $\{B_1,A_2,B_3\}$, $\{B_1,B_2,A_3\}$, and $\{B_1,B_2,B_3\}$. There are 8 such groups.

18. (A) Tuesday

Tuesday	2 up and out
Monday	2 up 1 down = 1 up
Sunday	2 up 1 down = 1 up
Saturday	2 up 1 down = 1 up
Friday	2 up 1 down = 1 up
Thursday	2 up 1 down = 1 up
Wednesday	2 up 1 down = 1 up
Tuesday	2 up 1 down = 1 up
Monday	2 up 1 down = 1 up

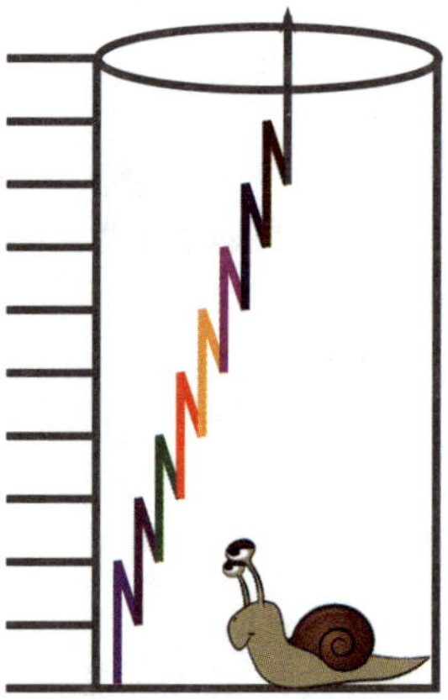

On the 9th day the snail climbs up from the level of 8 meters above the bottom of the well to the level of 10 meters from the bottom and gets out of the well.
This happens on Tuesday.

19. (C) 254361
The smallest 3-digit number John can create is 2 4 6 and Stan's largest 3-digit number is 5 3 1. Put them together in alternating order (John first) to see 254361 as the number formed by both of them.

20. (D) 80
The class spent $\$2 + \$3 + \$4 + \$5 = \$14$ per student to buy 4 different tickets for each student. For the whole class they spent \$280, so there were $280 \div 14 = 20$ students in the class. Since the school bought 4 tickets for each student, they bought $20 \times 4 = 80$ tickets.

21. (D) 10
$\frac{3}{4}$ of the capacity of one carton is the same as $\frac{3}{2}$ of the capacity of one glass, so $\frac{1}{4}$ of the capacity of one carton is the same as $\frac{1}{2}$ of the capacity of one glass, or the capacity of 1 cartoon is the same as the capacity of 2 glasses since $4 \times \frac{1}{4} = 1$ and $4 \times \frac{1}{2} = 2$. Therefore, juice from 5 cartons will fill 10 glasses.

22. (E) E

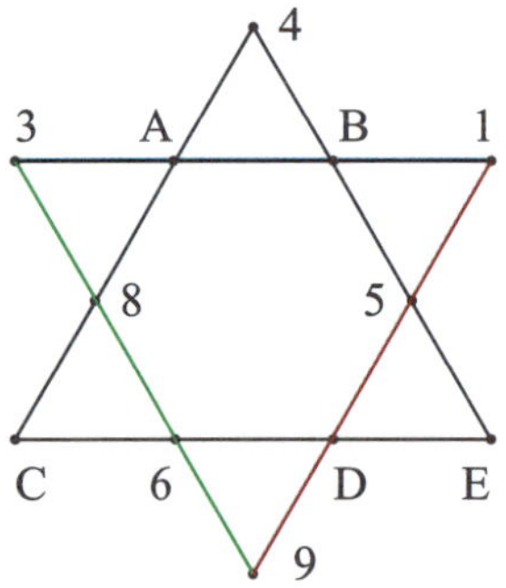

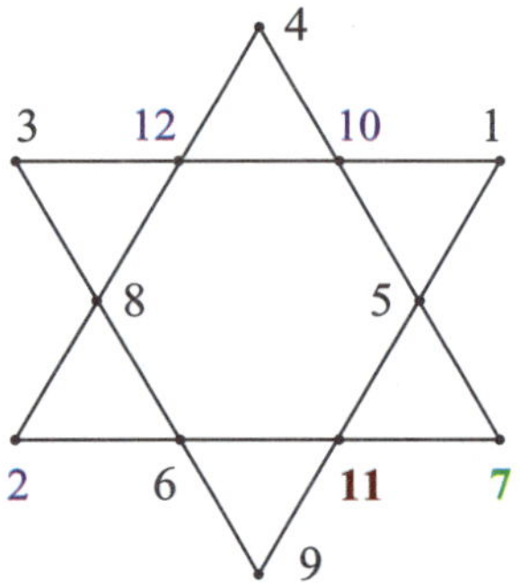

The numbers along the green segment add up to $3 + 8 + 6 + 9 = 26$, so 26 is the sum of numbers along any segment. Hence, the number under D is $26 - (1 + 5 + 9) = 11$. Also, the sum of numbers under C and E is 9 since $26 - (6 + 11) = 9$. The numbers still available are 2, 7, 10, and 12. To get 9 as the sum we can only use 2 and 7. Look at the segment with E and B. If the number under E were 2, then the number under B would be $26 - (2 + 5 + 4) = 15$; however, 15 is not among numbers from 1 to 12.
Thus, 7 is under E and we can find all the other numbers. The results are shown above.

23. (C) 5

The first ten perfect squares are 1, 4, 9, 16, 25, 36, 49, 64, 81, and 100. The sum of ones digits of these squares is $1 + 4 + 9 + 6 + 5 + 6 + 9 + 4 + 1 + 0 = 45$, so the ones digit of the sum $1^2 + 2^2 + 3^2 + 4^2 + 5^2 + 6^2 + 7^2 + 8^2 + 9^2 + 10^2$ is 5.

24. (C) 2, 7, 6, 3, 4, 5, 1

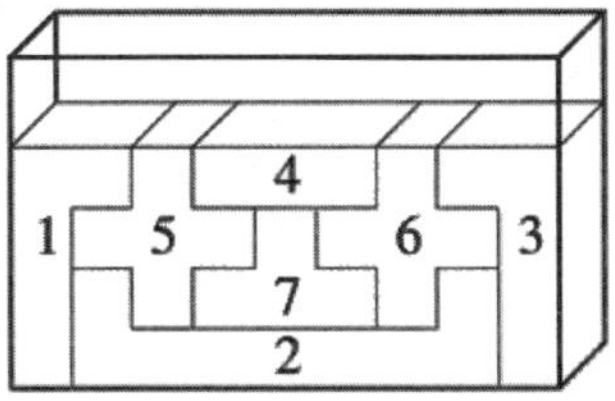

For orders (A) 2, 7, 5, 6, 4, 1, 3; (B) 2, 7, 5, 1, 6, 4, 3; (D) 2, 7, 6, 5, 3, 1, 4; and (E) 2, 7, 5, 1, 6, 3, 4 we can put all the blocs in the places they are shown in the box to the left. For (C), the blocks 2, 7, 6, and 3 can be easily placed in the box in that order. After that, placing block 4 before block 5 would not allow block 5 to fill the remaining empty spaces.

25. (B) 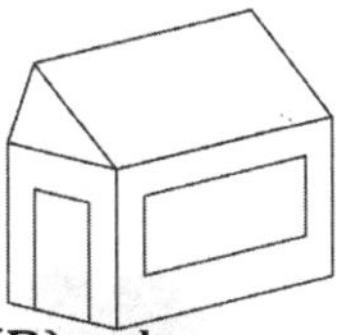

In (B) when we face the door of the house, the long window is on the right side.

(A)  (C) 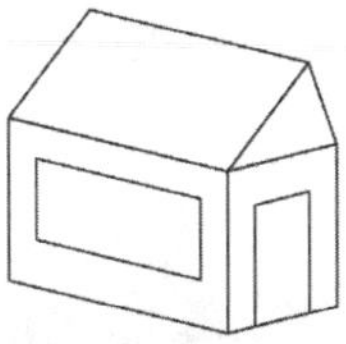 (D)  (E) 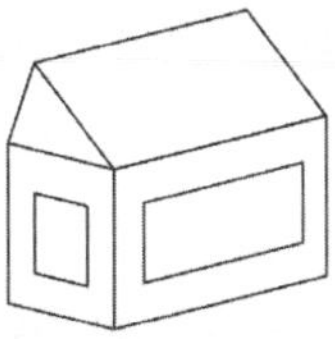

In (A) there are two windows on the right side, so the long window (not visible) is on the left side as shown in (C). The pictures of the house in (D) and (E) are consistent with (A) and (C), so the house Y is shown in (B).

26. (E) 5

There were four teams: team A earned 5 points, team B earned 3 points, team C earned 3 points, and team D earned 2 points. The four teams played 6 games: {A,B}, {A,C}, {A,D}, {B,C}, {B,D}, and {C,D}.

Each game yields either $3 + 0 = 3$ points or $1 + 1 = 2$ points. The total of all points is $5 + 3 + 3 + 2 = 13$. Neither 3 nor 2 are factors of 13, so at least one game yielded 3 points and at least one game yielded 2 points. The other $13 - (3 + 2) = 8$ points are distributed among 4 games. 8 as a sum of 4 numbers, each number being either 3 or 2, can be written only in one way: $8 = 2 + 2 + 2 + 2$. Thus, 5 games ended in a tie.

There is only one distribution of points. A and B were tied, A and C were tied, and A beat D. The other three games were tied.

27. (B) 4

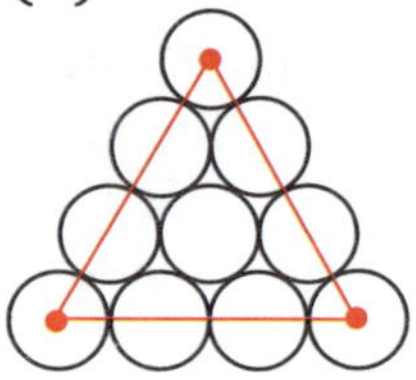
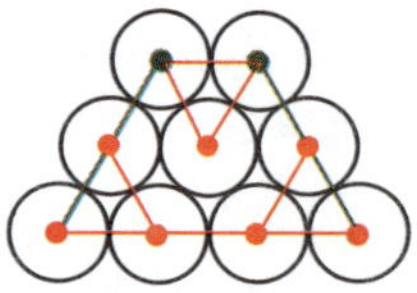

One of the corner coins with red centers must be removed. Suppose it is the top coin. The triangle with three orange sides and two triangles each with 3 sides of different colors have the same size but they do not share vertices. From each of these three triangles we have to remove at least one coin to prevent forming equilateral triangles. Hence, at least $1 + 3 = 4$ coins must be removed to prevent formation of equilateral triangles. It can be done as shown below to the left.

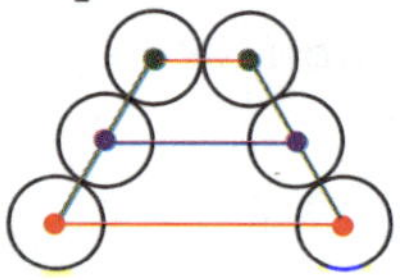

If d is a diameter of one coin, then d is the length of the orange segment, $2d$ is the length of the blue segment and $3d$ is the length of the red segment. The figure is symmetric with respect to the vertical line, so to see all triangles of different sizes pick 2 vertices from the right green segment (it can be done in 3 ways) and one vertex from the left green segment (there are 3 options), so $3 \times 3 = 9$ is the number of triangles of different sizes. For each of these triangles you can easily see that the three sides are not of the same length or not every of the three angles has 60°, so none of the triangles are equilateral.

A word of caution: Not every way of removing four coins guarantees that no equilateral triangles can be formed. For example, removing all corner coins with red centers would not work, as shown in the picture with the removed coins colored white.

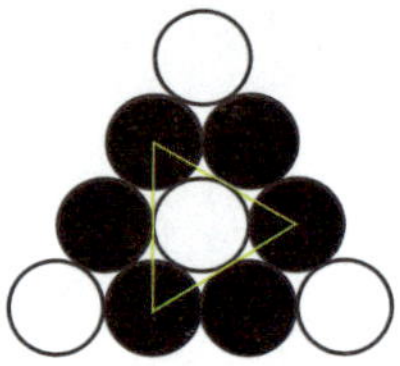

28. (C) 14

If Snow White divided 77 berries among 7 dwarfs equally, then each dwarf would get $77 \div 7 = 11$ berries. When lined up from shortest to tallest, the middle dwarf has 3 shorter dwarfs on one side and 3 taller dwarfs on his other side. The middle dwarf has 11 berries. His shorter neighbor has one berry less than the middle dwarf, and the taller neighbor has one more, so one has 10 berries and the other has 12 berries. The shorter neighbor of the dwarf with the 10 berries has two berries less than the middle dwarf and the taller neighbor of the dwarf with 12 berries has two more, so one has 9 and the other 13. The shortest dwarf has three berries less than the middle dwarf and the tallest dwarf has three more. Thus, the distribution of the 77 berries from the shortest to the tallest dwarf is 8, 9, 10, 11, 12, 13, 14. Hence, the tallest dwarf got 14 berries.

29. (D) 16

Any of the 4 teams can win the competition. Once the winner is selected, any of the 2 teams from the other matchup can be the second-place team. If the first and second places are known, then either one from the other 2 teams can be the third-placed team, so the number of all possible outcomes is $4 \times 2 \times 2 = 16$.

Here are all 16 outcomes: ACBD, ACDB, ADBC, ADCB; BCAD, BCDA, BDAC, BDCA; CADB, CABD, CBDA, CBAD; DACB, DABC, DBCA, DBAC.

Here is another way to find the possible outcomes: If A is the winner, then B can be only third or fourth and either C or D is second. Assuming that A places first in this competition, the possible outcomes are: ACBD, ACDB, ADBC and ADCB. There are 4 possible outcomes in the case when A is the winner. The same argument works when B, C, or D is the winner of the whole competition, so there are $4 \times 4 = 16$ possible outcomes.

30. (A) $\frac{1}{2}$

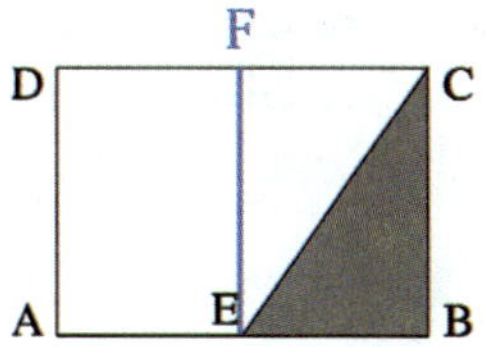

Draw a segment EF which is perpendicular to AB, so the triangles EBC and CFE are equal. Hence, the area of rectangle EBCF is $\frac{1}{4}+\frac{1}{4}=\frac{1}{2}$ of the area of rectangle ABCD. Thus, E is the midpoint of the segment AB, so the base of the shaded triangle is $\frac{1}{2}$ of the length of the base of the rectangle.

Solutions for Year 2000

1. (C) 16
To the class of 29 students add 3 new boys to equal the number of boys and girls. Then this new class has 32 students and half of them are girls. There are $(29 + 3) \div 2 = 16$ girls in the class.

2. (A) 3
$-11 - 2(-7) = -11 - (-14) = -11 + 14 = 3$

3. (D) $585 \times 3 \times 5$
There are 3 mice in each drawer, so there are (585×3) mice. Each mouse has 5 little mouse pups, so there are $(585 \times 3) \times 5 = 585 \times 3 \times 5$ little mouse pups in the chest.

4. (E) 402
Let m be the middle number of the five consecutive numbers. Then the five numbers are $m-2$, $m-1$, m, $m+1$, $m+2$. Their sum is $5m$, so $5m = 2000$ and $m = 400$.
The greatest of the five numbers is $m + 2 = 400 + 2 = 402$.

5. (D) 29 km
30 minutes = 3×10 minutes and the train moves 9 km in 10 minutes, so in 30 minutes the train moves 3×9 km = 27 km. It moves towards the station, so after 30 minutes the distance will decrease from 56 km to 56 km – 27 km = 29 km.

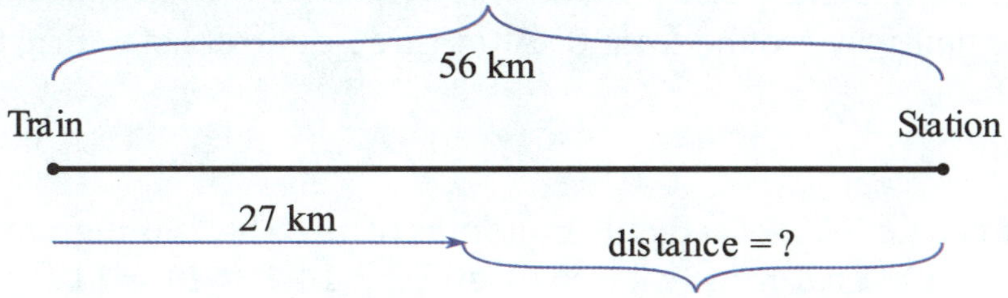

6. (E) 9:45

If the images of a wall clock and its mirror reflection are put on one clock, then the images are symmetric with respect to the vertical line passing through the center of the clock.

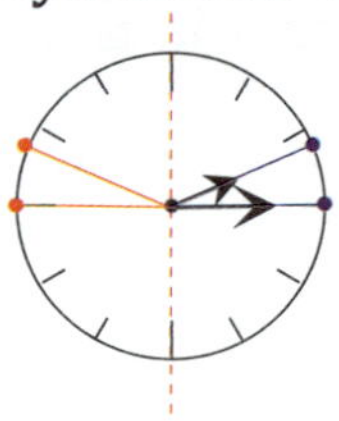

The hour and the minute hands of the mirror reflection point to the blue dots. The symmetry with respect to the vertical axis is applied and the results are shown in orange.

Thus, the minute hand of the actual wall clock points towards 9 indicating 45 minutes past the hour, and the hour hand points somewhere between 9 and 10 (closer to 10). Hence, the wall clock shows 9:45.

7. (D) 4 twos and 3 fives

$2000 = 2 \times 1000 = 2 \times 10^3 = 2 \times (2 \times 5)^3 = 2 \times 2^3 \times 5^3 = 2^4 \times 5^3$, so among the prime factors of 2000 there are 4 twos and 3 fives.

8. (B) 240 cm

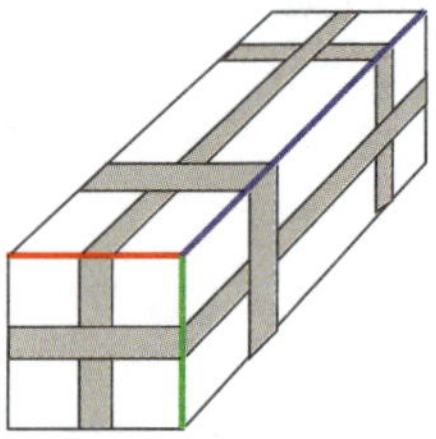

Three faces of the box are visible. Each of them has a number of pieces of the ribbon. Add the lengths of all visible pieces and then double the sum (to account for the nonvisible faces) to get the length of the whole ribbon. There are 3 types of visible pieces: 3 pieces that are parallel to the red edge, 3 pieces that are parallel to the green edge, and 2 pieces that are parallel to the blue edge. The dimensions of the box are 10 cm × 10 cm × 30 cm, so the 6 short pieces have lengths of 10 cm each and the 2 long pieces have a length of 30 cm each. $6 \times 10 + 2 \times 30 = 120$, so the combined length of the visible pieces is 120 cm. The length of the whole ribbon is 2×120 cm $= 240$ cm.

9. (E) 7

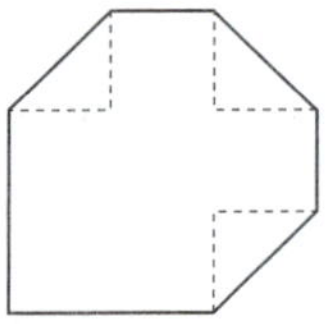

One folded corner makes 2 vertices. Two folded corners make 4 vertices and three folded corners make 6 vertices. Thus, the final figure with 6 vertices by the three folded corners and one unfolded corner has 7 vertices.

10. (E) E

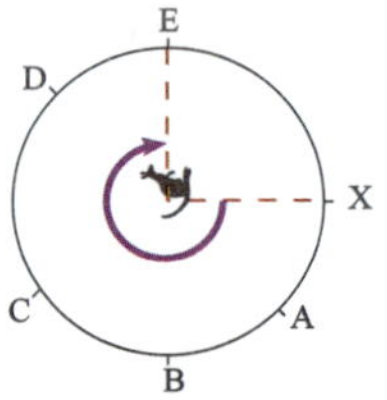

A rotation around a whole circle is 360°, so 270° is three-quarters of the way around. Going clockwise, the letter B is 90° or one-quarter turn from X, and then E is 180° or half-way around from B. After three-quarters of a turn, the kangaroo faces E.

11. (B) 7

Multiples of both 2 and 7 are exactly the multiples of 14. The two-digit multiples of 14 are $1 \times 14, 2 \times 14, 3 \times 14, 4 \times 14, 5 \times 14, 6 \times 14$, and 7×14. $8 \times 14 = 112$, so it is not a two-digit number. There are 7 such numbers.

12. (E) E

Eliminate all subtractions from $A - 1 = B + 2 = C - 3 = D + 4 = E - 5$ by adding 5 to each expression. Then $A + 4 = B + 7 = C + 2 = D + 9 = E$, so E is greater than any other number. Hence, E is the greatest number among the five numbers.

13. (D) 55
The figure in the picture has 5 steps and also 5 small squares in the bottom row. For this figure the number of small squares is $1 + 2 + 3 + 4 + 5$.
For a figure with 10 steps the number of small squares is:
$1 + 2 + 3 + 4 + 5 + 6 + 7 + 8 + 9 + 10 =$
$= (1 + 10) + (2 + 9) + (3 + 8) + (4 + 7) + (5 + 6) = 5 \times 11 = 55$.

14. (B) 166 hours and 40 minutes
$1{,}000{,}000 = 10{,}000 \times 100$, so it will take 10,000 minutes to print one million forms.
10,000 minutes $= 166 \times 60$ minutes $+ 40$ minutes $=$ 166 hours and 40 minutes.

15. (C) Mr. Jack
Plots (B), (D), and (E) have the same fence length as plot (A). For each of the four plots except plot (A), we can match segments of the solid line with the segments of the dashed line as indicated below by letters u, d, *l*, r. At a given segment the letters indicate: move it up, down, left, or right, respectively.

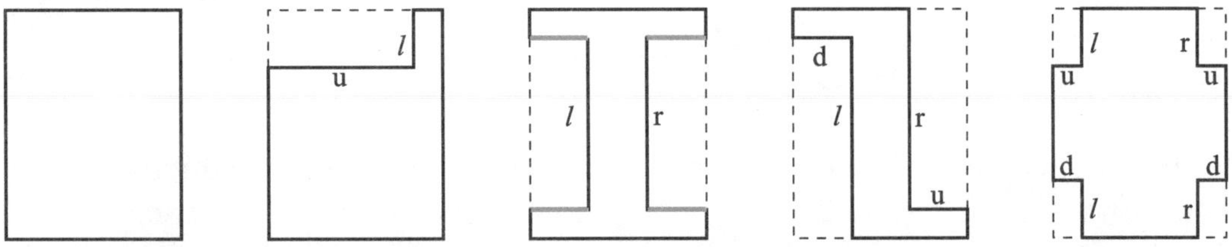

(A) (B) (C) (D) (E)

Only for figure (C), Mr. Jack's plot, there are solid line segments (the four short horizontal segments marked in green) without dashed matches, so the length of Mr. Jack's fence is greater (by the four short horizontal segments) than any other fence.

16. (E) decrease by 7
Decrease number a by 5 to get $a - 5$. Increase number b by 2 to get $b + 2$. Compute the difference between the new numbers: $(a - 5) - (b + 2) = a - 5 - b - 2 = (a - b) - 7$.
Thus, the difference $a - b$ is decreased by 7. A numerical value of $a - b$ is not needed for this solution.

17. (D) 60 days
We need the least common multiple of 1, 2, 3, 4, 5, and 6. 30 is a multiple of 1, 2, 3, 5, and 6 but it is not a multiple of 4. We need 60 days to see John, Karl, Stan, Adam, Paul, and Peter together again at the lab.

18. (E) 10
There are three triangles with the base 1, two triangles with the base 2, and one triangle with the base 3. 2 is the common height for all these triangles, so the sum of the areas is
$\frac{1}{2} \times 2 \times (1 + 1 + 1 + 2 + 2 + 3) = 10$.

19. (D) 4, 9, 2, 5
The smallest 3-digit numbers start with 1. This eliminates digits 4, 9, and 2 from 4921508, so we are left with 1508. Next we have to eliminate 5, 0, or 8 to get a 3-digit number starting with 1. The smallest among them is 108, so we need to remove the digit 5. Thus, 4, 9, 2, and 5 must be removed.

20. (D) 10
The first digit can be 1, 2, or 3. If it is 3, then the only acceptable 4-digit number is 3000.
If the first digit is 2, then the acceptable numbers are 2100, 2010, and 2001.
If the first digit is 1, then the acceptable numbers are 1200, 1020, 1002 and the numbers 1110, 1101, and 1011.
Together, there are $1 + 3 + 6 = 10$ four–digit numbers with a sum of digits equal to 3.

21. (D) 4
If 96 participants are divided into equal groups, then the number of participants in each group is a divisor of 96. We list all divisors of 96 in pairs: 1, 96; 2, 48; 3, 32; 4, 24; 6, 16; and 8, 12. Among the divisors only 6, 8, 12, and 16 are between 5 and 20. There are 4 of them, so the division into equal groups can be done in 4 ways.

22. (D) 70 g
Keeping the scale still in balance we can remove two cubes and one cylinder from each side of the scale. Thus, one cube is balanced by 20 g + one cylinder. Five cubes are balanced by 100 g + five cylinders. Together on the scale there are 5 cubes and 3 cylinders. They weigh 500 g but the weight of five cubes is 100 g + the weight of five cylinders, so
500 g = 100 g + the weight of five cylinders + the weight of three cylinders. Therefore,
400 g = the weight of eight cylinders. One <u>cylinder</u> weighs 50 g and one cube weighs 20 g + one cylinder, so one cube weighs 20 g + 50 g, which is 70 g.

23. (B) It decreased by 1%.
Let h and w be the sides of a given rectangle. If we increase h by 10% and decrease w by 10%, then the area of the new rectangle is $(1.10 \times h) \times (.90 \times w) = .99 \times (h \times w)$, which is .01, or 1%, less than $h \times w$. Since $h \times w$ is the area of the given rectangle, the area decreased by 1%.

24. (B) 9
A rectangle (shown in black) has the width twice the size of the height. We divide each height into two equal intervals and each width into four equal intervals. All these 12 intervals have the same length, which is half of the height. The length of two such intervals is exactly the length of the height. The red lines are the axes of symmetry for the black rectangle connecting the midpoints of the opposite sides. An endpoint of the vertical red line is the midpoint of the segment with blue endpoints, so the length of this segment is equal to the length of the height. The whole rectangle is divided into 16 equal right isosceles triangles. The shaded area consists of 8 of them, so the shaded area is half of the area of the big rectangle. The original dimensions of the big rectangle are 3×6, so the shaded area is $(3 \times 6) \div 2 = 9$.

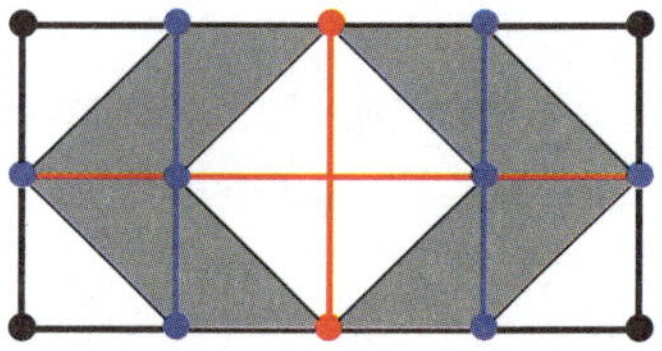

25. (A) 30

The mother needs sixty 3-meter jumps to cover 180 meters. One of her jumps takes 1 second, so after 60 seconds she reaches the eucalyptus. During these 60 seconds the little kangaroo is making 120 jumps since it needs only one-half of a second for 1 jump. Its one jump is one meter long, so after the first 60 seconds the little kangaroo is 120 meters away from the start and at the distance of (180 – 120) meters = 60 meters from the eucalyptus. The mother is waiting while the little kangaroo is covering the last 60 meters. It needed 60 seconds to cover 120 meters, so it needs 30 seconds to cover the last 60 meters. These 30 seconds is how long the mother will wait at the eucalyptus for the little kangaroo.

26. (A) 15°

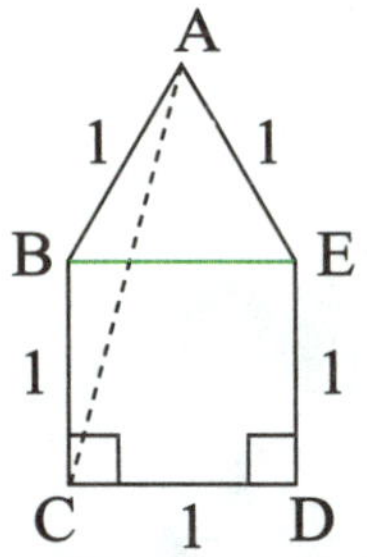

BE = 1, so △BEA is an equilateral triangle and ∠ABE = 60°.
∠EBC = 90°, so ∠ABC = ∠ABE + ∠EBC = 60° + 90° = 150°.
△CAB is an isosceles triangle, so ∠BAC = (180° – 150°) ÷ 2 = 15°.

27. (D) 13

Below the counterweight numbers are in **bold**.
By using only one of the three weights we can get 1 = **1**, 3 = **3**, and 9 = **9**.
Placing two weights on one side and none on the other side gives us 1 + 3 = **4**, 1 + 9 = **10**, and 3 + 9 = **12**.
Using two weights but not placing both on one side gives us 3 – 1 = **2**, 9 – 1 = **8**, and 9 – 3 = **6**.
Placing all three weights on one side gives us 1 + 3 + 9 = **13**.
Using all three weights, but not placing all of them on one side, gives us 9 – (1 + 3) = **5**, (1 + 9) – 3 = **7**, and (3 + 9) – 1 = **11**. The numbers **1, 3, 9, 4, 10, 12, 2, 8, 6, 13, 5, 7,** and **11** are different, so 13 counterweights can be found.
Below is a visual solution. The 1 kg weight is in yellow, the 3 kg weight is in green, the 9 kg weight is in purple, and the counterweight is in blue.

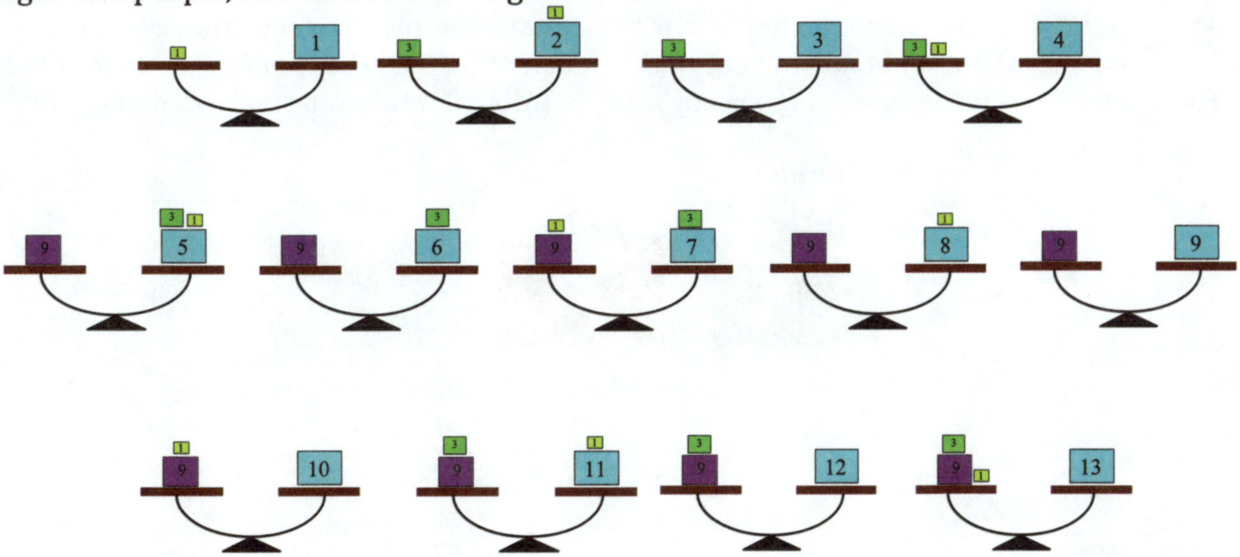

28. (E) 64

A cube of the size 4 cm × 4 cm × 4 cm consists of 8 smaller cubes of the size 2 cm × 2 cm × 2 cm (see the picture). Since one smaller cube uses 8 grams of play dough, 8 × 8 grams = 64 grams of play dough is needed to make the 4 cm × 4 cm × 4 cm cube.

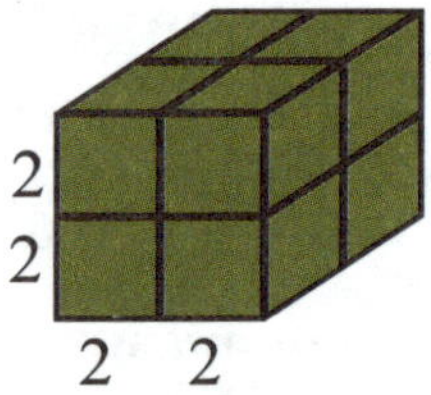

29. (D) 10

Let G stand for "green" and Y for "yellow." Starting with the front of caterpillar, we can color the five spherical parts by choosing where the 2 green parts are placed (the other 3 parts are yellow).

We can list the possibilities in alphabetical order as follows.
GGYYY; GYGYY, GYYGY, and GYYYG; YGGYY, YGYGY, and YGYYG; YYGGY, YYGYG; and YYYGG.
Together, there are 10 options for this type of caterpillar.
Below is a visual solution.

30. (A) in the red box

The red box is to the right of the shell, so the first box is not red.
The bead is to the right of the red box, so the last box is not red. Hence, the middle box is red.
The green box is to the left of the blue box but neither is in the middle, so the green box must be the first one and the blue box must be the last one.
Again, *the red box is to the right of the shell* and the red box is in the middle, so the shell is in the green box.
Similarly, *the bead is to the right of the red box* and the red box is in the middle, so the bead is in the blue box.
Therefore, the coin must be in the red box.
At this point, the colors and the items in the boxes are completely determined.
So far our reasoning is not making use of the information that *the coin is to the left of the bead.*
Fortunately, this is the case, so (E) is not a valid option and the solution is shown below.

Solutions for Year 2002

1. (B) 2323
The first and the last digits are different.

2. (C) 

Below four lines are identified as segments of the sketch.

(A)  (B)  (C)  (D)  (E) 

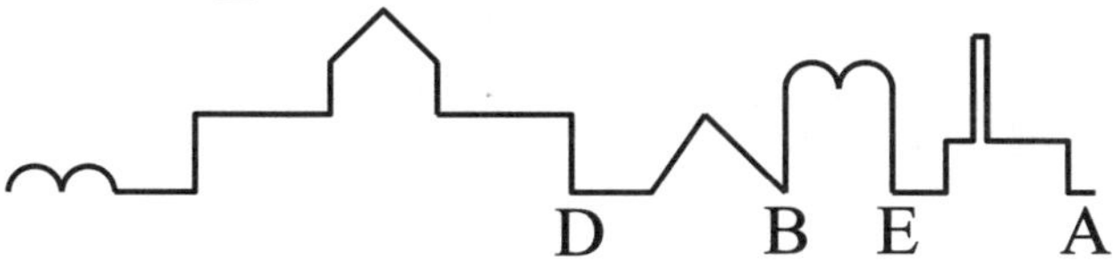

Line (C) does not belong to the sketch.

3. (D) 5
The two boys are brothers to all three daughters. $3 + 2 = 5$, so Mr. and Mrs. Kowalski have 5 children, 3 girls and 2 boys.

4. (E) 231
$3^2 = 3 \times (2 + 1)$
For other numbers: $9^2 > 3 \times (1 + 2)$, $4^2 < 3 \times (7 + 1)$, $8^2 > 3 \times (3 + 5)$, and $3^2 < 3 \times (1 + 8)$.

5. (C) 2^{2003}
We add the exponents when multiplying powers of the same base, so $2^2 \times 2^{2000} \times 2 =$
$= 2^2 \times 2^{2000} \times 2^1 = 2^{2+2000+1} = 2^{2003}$.

6. (D) 
There are 4 black hearts on this string of 6 hearts and $4 = \frac{2}{3} \times 6$. For the other strings the equation $b = \frac{2}{3} \times a$ ($b =$ the number of black hearts, $a =$ the number of all hearts) is false.

(A) 
$b = 2, a = 5$

(B) 
$b = 6, a = 10$

(C) 
$b = 2, a = 6$

(E) 
$b = 6, a = 12$

7. (D) $10{,}000 \times 100 \div 10$
When multiplying by 10 move the decimal point to the right, and when multiplying by a decimal move the decimal point to the left. When dividing, move the decimal point the other direction.

(A) $10 \times 0.001 \times 100 = 1000 \times 0.001 = 1$
(B) $0.01 \div 100 = .0001$
(C) $100 \div 0.01 = 10{,}000$
(D) $10{,}000 \times 100 \div 10 = 100{,}000$
(E) $0.1 \times 0.01 \times 10{,}000 = 10$

Of these, the greatest number is 100,000, answer (D).

8. (C) 58

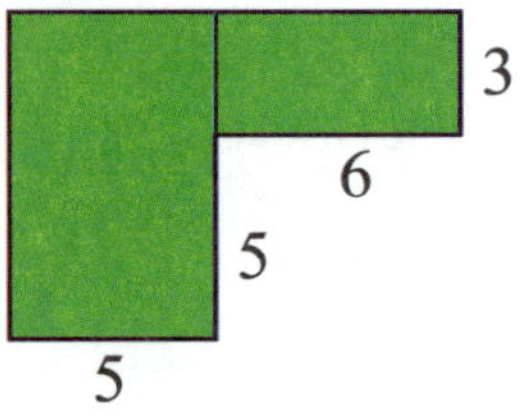

The black vertical line divides the figure into two rectangles. The left rectangle has the dimensions $(5 + 3) \times 5$ and the right rectangle has the dimensions 3×6, so the combined area is $(8 \times 5) + (3 \times 6) = 58$.

9. (E) 1,250 cm^2

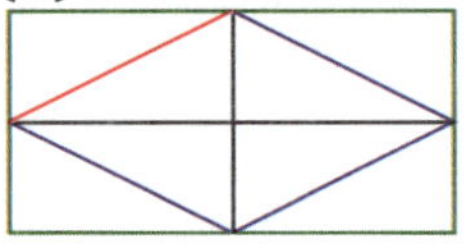

A rectangle with all six pairs of midpoints of its sides connected is shown. The rectangle is divided into 8 equal triangles, so the area of each triangle is $\frac{1}{8} \times 1\text{ m}^2$. $1\text{ m} = 100\text{ cm}$, so $1\text{ m}^2 = 100\text{ cm} \times 100\text{ cm} = 10{,}000\text{ cm}^2$ and $\frac{1}{8} \times 10{,}000\text{ cm}^2 = 1{,}250\text{ cm}^2$, so each triangle (including the cut-off one) has an area of $1{,}250\text{ cm}^2$.

10. (B) 885
The smallest three-digit number with all different digits is 102 and the greatest three-digit number with all different digits is 987. The difference between them is 885.

11. (C) 64 m

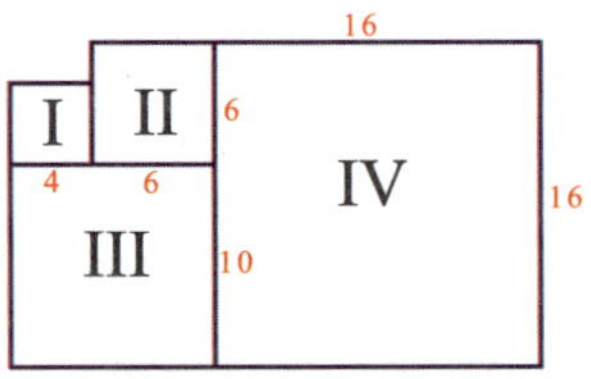

Each side of the square I has the length of $16\text{ m} \div 4 = 4\text{ m}$.
Each side of the square II has the length of $24\text{ m} \div 4 = 6\text{ m}$.
Hence, the top side of the square III, and any other side of this square, has a length of $(4\text{ m} + 6\text{ m}) = 10\text{ m}$. The left side of the square IV has a length of $(6\text{ m} + 10\text{ m}) = 16\text{ m}$. Therefore, the perimeter of the square IV is $4 \times 16\text{ m}$, which is 64 m.

12. (A) 85
$64 = 16 \times 4$, so there are 16 groups of 4 plates each. From each group we can form 4 medals and 1 new plate, so we can form 16×4 medals $=$ **64** medals and 16 new plates. Put the 64 medals aside. $16 = 4 \times 4$, so there are 4 groups of 4 new plates each. From each group we can form 4 medals and one extra plate, so we can form 4×4 medals $=$ **16** medals and 4 extra plates. From the last 4 plates we can form **4** medals and 1 plate. From the last plate **1** medal can be cut out, so the largest number of medals that can be formed from 64 plates is $64 + 16 + 4 + 1 = 85$.

13. (C) 7

The area of rectangle ABCD is 24. If we remove the three right triangles, we will get the area of triangle ALM. The area of any right triangle is half of the product of the lengths of its legs, so the area of triangle ABL is $(6 \times 2) \div 2 = 6$, the area of triangle LCM is $(2 \times 1) \div 2 = 1$, and the area of triangle MDA is $(5 \times 4) \div 2 = 10$. $6 + 1 + 10 = 17$ is the total area removed from the rectangle, so the area of triangle ALM is $24 - 17 = 7$.

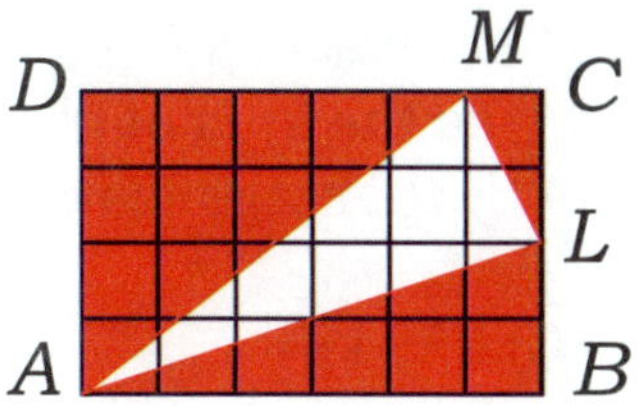

14. (E) 22 and 23

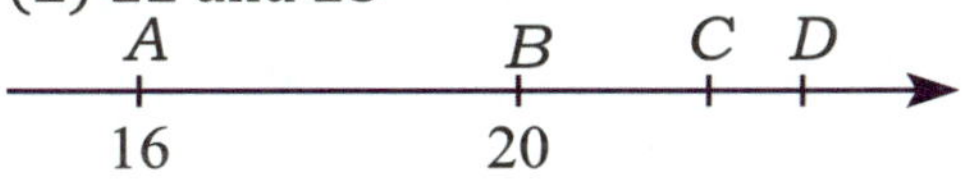

$|AB| = (20 - 16) = 4$, $|BC| = 4 \div 2 = 2$ and $|CD| = 2 \div 2 = 1$. C is to the right of B, so the coordinate of C is $20 + 2 = 22$. D is to the right of C, so the coordinate of D is $22 + 1 = 23$.

15. (E) 0

In any triangle, the sum of the lengths of any two sides is strictly greater than the length of the third side. If m = the length of the medium side and ℓ = the length of the longest side, then $m < \ell$. If both m and ℓ are natural numbers, then $(m + 1) \leq \ell$ since $m + 1$ is the smallest **natural** number greater than m. If 1 is the shortest side of a triangle (with the other two sides being m and ℓ), then $m + 1 > \ell$ but $(m + 1) \leq \ell$, so the stick of the length 1 cannot be a side of any triangle built from three of the given sticks.

16. (C) 8

A convex angle is an angle greater than 0° but smaller than 180°.
$50° + 30° + 20° + 10° = 110° < 180°$, so we can pick any two rays to have a convex angle. There are 5 rays and we can pick two at a time in 10 different ways. The measures of the 10 angles are: 50°, $50° + 30° = 80°$, $50° + 30° + 20° = 100°$, $50° + 30° + 20° + 10° = 110°$; 30°, $30° + 20° = 50°$, $30° + 20° + 10° = 60°$; 20°, $20° + 10° = 30°$; and 10°. Among them there are 8 different measures.

17. (B) 9

A number is a multiple of 25 if it ends with 00, 25, 50, or 75. Since 2 is not one of the given digits, we are left with three possibilities: 00, 50, or 75. The first digit can only be 3, 5, or 7, so there are 3×3 numbers that are 3-digit numbers divisible by 25 and use only digits 0, 3, 5, 7. Here is a complete list of such numbers: 300, 350, 375; 500, 550, 575; 700, 750, 775.

18. (A) Oliver has a dog.

Nate, who has a pet with fur but doesn't like cats, has to have a dog. Because Nate owns a dog, the sentence, "Oliver has a dog," is not true. Actually, Oliver owns a cat. The other sentences are true.

19. (A) on Monday

If the day after tomorrow is Thursday, then tomorrow is Wednesday, and today is Tuesday. It is one day after Johnny's birthday, so he had his birthday on Monday.

20. (C) 23
The area of triangle *BCE* is the area of triangle *ABC* minus the area of triangle *ABE*. $15 - 4$ is 11, so the area of triangle *BCE* is 11. Similarly, the area of triangle *EDA* is $12 - 4 = 8$. The area of the pentagon is $4 + 11 + 8 = 23$.

21. (C) 239 kg
Each boy was weighed with each of the other 4 boys, so each boy was weighed 4 times. Add the ten results and divide by 4 to see the total weight of all five boys.
$90 + 92 + 93 + 94 + 95 + 96 + 97 + 98 + 100 + 101 =$
$(90 + 100) + (92 + 98) + (93 + 97) + (94 + 96) + (95 + 101) = 4 \times 190 + 196 = 956$.
$956 \div 4 = 239$, so the total weight of all five boys equals 239 kg.

22. (B) $S_3 < S_1 = S_2 = S_4$
Let S be the area of each square.

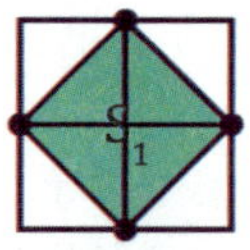

The whole square is divided into 8 equal triangles and 4 of them are shaded, so $S_1 = \frac{1}{2}S$.

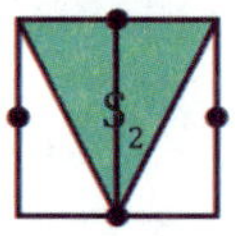

The whole square is divided into 4 equal triangles and 2 of them are shaded, so $S_2 = \frac{1}{2}S$.

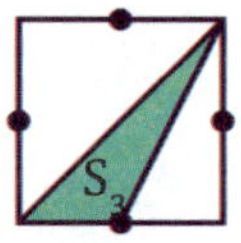

The diagonal is dividing the square into 2 equal parts. One part consists of the shaded triangle and the right triangle at the lower right corner. Both triangles have the same height and their bases have the same length (half of the square side), so they have the same area which is $\frac{1}{4}S$. Hence, $S_3 = \frac{1}{4}S$.

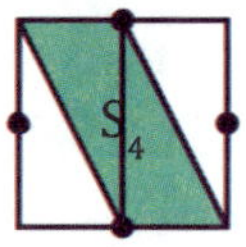

The whole square is divided into 4 equal triangles and 2 of them are shaded, so $S_4 = \frac{1}{2}S$.

S_3 has an area that is smaller and all the other areas are the same, so $S_3 < S_1 = S_2 = S_4$ is the only true statement.

23. (D) 39
The multiples of 3 between 1 and 100 are: $1 \times 3, 2 \times 3, 3 \times 3, 4 \times 3, \ldots, 33 \times 3$.
There are 33 of them. Numbers which are not multiples of 3 but have the last digit 3 are: 13, 23, 43, 53, 73, and 83. There are 6 of them. Together, there are $33 + 6 = 39$ of them, so you will clap your hands 39 times.

24. (A) 24

The cyclist went up and down the same hill, so the distance travelled up and down is the same. The formula for calculating distance travelled is speed multiplied by time.
The time it took the cyclist to travel down the hill was t minutes, and the time it took to go up was $t+16$ minutes. Using the formula for distance travelled, the distance of going uphill is 12km/h times $t+16$ minutes, and the distance going downhill is 20km/h $\times$ t minutes. Since the distance of uphill and downhill are the same, $12 \times (t+16) = 20t$
Solving, we get

$$12 \times t + 12 \times 16 = 20t$$
$$12t + 192 = 20t$$
$$192 = 8t$$
$$t = 24$$

This mean that it takes the cyclist 24 minutes to travel down the hill.

25. (A) $P < S < Q$

The weight of 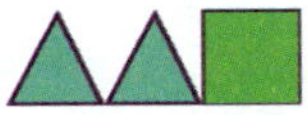 is P and the weight of  is Q. Each contains one square and $P < Q$, so we can remove each square to see that 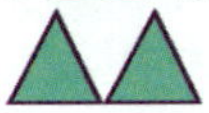 weigh less than . Hence,  weighs less than , or  < .

Add 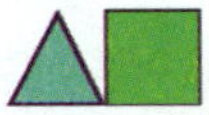 to both sides of the last inequality. Then  < , so $P < S$.
Use  <  again. Add 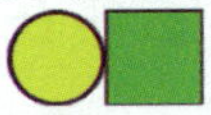 to both sides to get  < . Hence, $S < Q$.
In conclusion, $P < S < Q$. The other statements are false since each of them contradicts the true statement that $P < S < Q < R$.

26. (A) 23

With 14 gray marbles, 8 white marbles, and 6 black marbles it may happen that among the first 22 marbles $(22 = 14 + 8)$ taken out there are no black marbles. To make sure that among randomly selected marbles at least one is black, Ada must take out at least 23 marbles. Similarly, to have at least one white marble among randomly selected marbles Ada must take out at least 21 marbles $(21 = 14 + 6 + 1)$. To have at least one gray marble among randomly selected marbles Ada must take out at least 15 marbles $(15 = 8 + 6 + 1)$. Hence, to have at least one marble of each color Ada must take out at least 23 marbles.

27. (A) $\frac{1}{5}$

On the first day, the virus destroyed $\frac{1}{2}$ of the memory, so there was $\frac{1}{2}$ of the original memory left at that point. On the second day, it destroyed $\frac{1}{2} \times \frac{1}{3} = \frac{1}{6}$ of the original memory, so there was $\frac{1}{2} - \frac{1}{6} = \frac{3}{6} - \frac{1}{6} = \frac{2}{6} = \frac{1}{3}$ of the original memory left. On the third day, the virus destroyed $\frac{1}{3} \times \frac{1}{4} = \frac{1}{12}$ of the original memory, so there was $\frac{1}{3} - \frac{1}{12} = \frac{4}{12} - \frac{1}{12} = \frac{3}{12} = \frac{1}{4}$ of the original memory left. On the fourth day, the virus destroyed $\frac{1}{4} \times \frac{1}{5} = \frac{1}{20}$ of the original memory, so there was $\frac{1}{4} - \frac{1}{20} = \frac{5}{20} - \frac{1}{20} = \frac{4}{20} = \frac{1}{5}$ of all the computer's memory left after those four days.

28. (B) 10

If xyz is a given 3-digit number, then $1 \le x \le 9$ and $0 \le y, z \le 9$. The sum $s = x + y + z$ is a number between $1 + 0 + 0 = 1$ and $9 + 9 + 9 = 27$. We add the digits of s searching for the largest result. If s is a one-digit number, then the sum of its digits is s. The largest sum among them is 9. Next we have $s = 10, 11, 12, ..., 19$, where the largest sum is $1 + 9 = 10$. In the last group of 20, 21, 22, ..., 27, the largest sum is $2 + 7 = 9$.
10 is the greatest value for the sum of the digits of s. It happens any time $s = x + y + z = 19$. There are 45 such numbers. We may list them in a systematic way.
199; 289, 298; 379, 388, 397; 469, 478, 487, 496; 559, 568, 577, 586, 595; 649, 658, 667, 676, 685, 694; 739, 748, 757, 766, 775, 784, 793; 829, 838, 847, 856, 865, 874, 883, 892; 919, 928, 937, 946, 955, 964, 973, 982, 991.

29. (C) 91

In the first stage there were 8 groups each with 4 players. Since each of the players played a game with each of three other players, each group played 6 games, so in the 8 groups the players played $8 \times 6 = 48$ games.
48 is the number of games in the first stage.
The second stage started with 16 players (2 from each of the initial 8 groups) so there were 4 groups with 4 players each. Again, each group of 4 players played 6 games, so in the 4 groups the players played $4 \times 6 = 24$ games.
24 is the number of games in the second stage.
The third stage started with 8 players (2 from each of the previous 4 groups) so there were 2 groups with 4 players each. Each group played 6 games, so in the 2 groups the players played $2 \times 6 = 12$ games.
12 is the number of games in the third stage.
The fourth stage started with 4 players (2 from each of the last 2 groups).
They played **6 games** and the two top players played **1 more game**, so during the whole competition $48 + 24 + 12 + 6 + 1 = (48 + 12) + (24 + 6) + 1 = 91$ games were played.

30. (A) 123

The first row and the third one have the same number of hexagons but the middle row has one hexagon less. Together there are 32 hexagons, so the first and the third row have 11 hexagons each and the middle row has 10 hexagons. It is easy to count all non-vertical matches in the first and the third row since these matches do not overlap. There are $11 \times 4 + 11 \times 4 = 88$ **non-vertical** matches in all **three** rows since the non-vertical matches of the middle row have been already counted.
The first row has 11 hexagons, so it has 12 vertical matches. The same is true for the third row. The second row has 10 hexagons, so it has 11 vertical matches.
Together, $88 + 12 + 12 + 11 = 123$, so 123 matches were used to make this net.

Solutions for Year 2004

1. (C) 909

$1000 - 100 + 10 - 1 = 900 + 10 - 1 = 910 - 1 = 909$

2. (C) 3

1		x	2
4	1		
	3		
	2		

4 is needed in the second column of the table. After that, a 3 is needed in the first row, so Caroline should put 3 in the square marked with x. There are two complete solutions, shown to the right.

1	4	3	2
4	1	2	3
2	3	1	4
3	2	4	1

1	4	3	2
4	1	2	3
2	3	4	1
3	2	1	4

3. (E) 2000×800

$(10 \times 100) \times (20 \times 80) = (20 \times 100) \times (10 \times 80) = 2000 \times 800$. The answer has 5 zeros. Note that in these products the zeros will be the same as all the zeros in both factors. None of the options other than (E) has 5 zeros.

4. (E) more than 90 hours

360,000 seconds = 6,000 minutes = 100 hours

5. (E) 2003

$20042003 = 10000 \times 2004 + 2003$ and 2003 is a non-negative integer less than 2004, so 2003 is the remainder when dividing 20042003 by 2004.

6. (D) 

To find a match for the sheet, we have to rotate the original figure by 90° either to the right (clockwise) or to the left (counterclockwise). We are looking to have a black square in each spot on either the original sheet or on the answer choice.

Clockwise rotation gives us , which is not a match for any of the five options.

Counterclockwise rotation gives us . Combined with answer (D), it makes a completely black square. Figure (D) is the only match.

7. (D) 8

$2004 = 2 \times 2 \times 3 \times 167$. 3 is a factor, $2 \times 2 = 4$ is a factor, $2 \times 3 = 6$ is a factor, and $2 \times 2 \times 3 = 12$ is a factor. 8 is not a factor of 2004.

8. (B) 28

The parents ate $73 - 12 = 61$ carrots since the son ate 12 carrots. Since the father ate 5 more carrots than the mother, the remaining $61 - 5 = 56$ carrots were split evenly between the father and the mother. Thus, the mother ate $56 \div 2 = 28$ carrots that week.

9. (C) 2400 m

There are 9 bus stops, so there are 8 equal segments between consecutive bus stops. Each segment has the length of $600 \text{ m} \div 2 = 300 \text{ m}$ since the distance from the 1st to the 3rd bus stop consists of two equal segments. The bus route is $8 \times 300 \text{ m} = 2400$ meters long.

10. (D) 3

$1-(2-(3-(4-5))) = 1-(2-(3-(-1))) = 1-(2-(3+1)) = 1-(2-4) = 1-(-2) = 1+2 = 3$

11. (D) 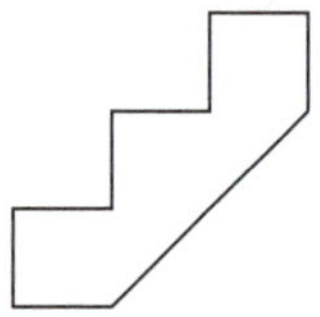

The original pieces and their 3 clockwise rotations (90°, 180°, and 270°) are shown in different colors:

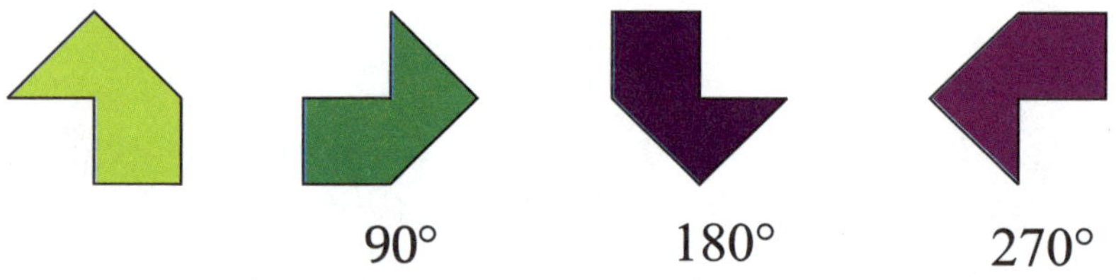

These colors show how to make four of the five figures using two pieces of different colors for each figure.

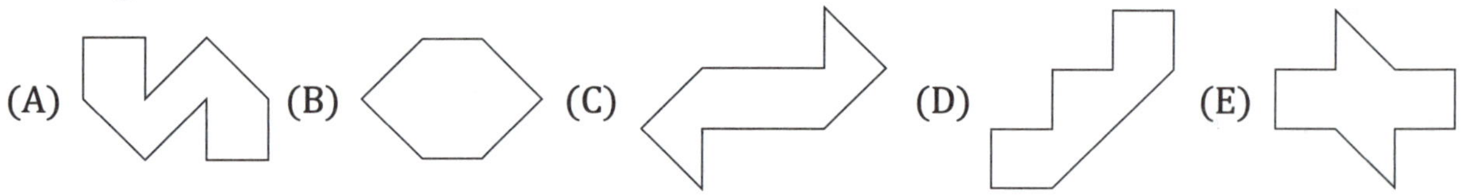

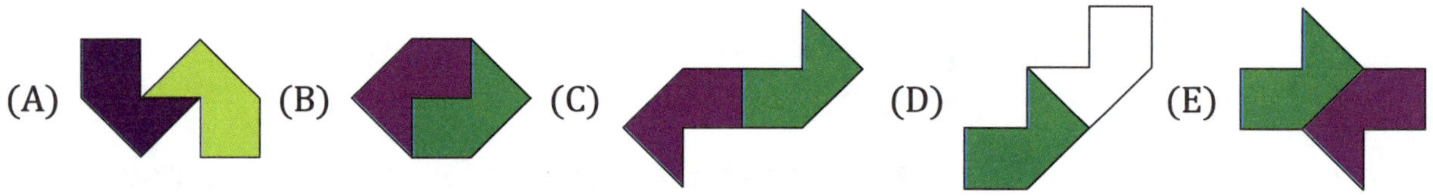

No rotation of the original piece fits the white area of (D). To complete (D), we would have to turn over one of the pieces.

12. (E) 32

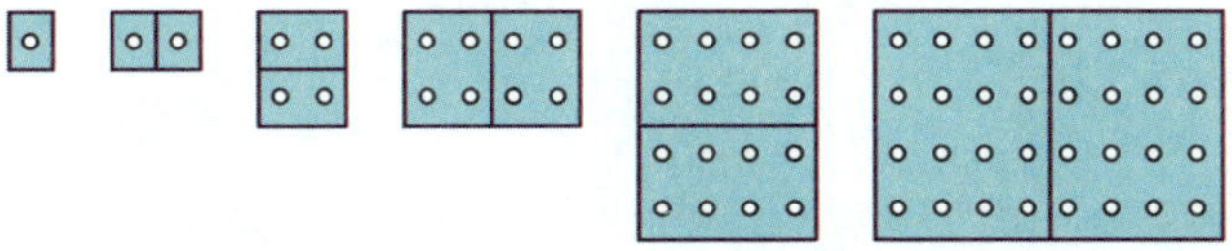

Working backwards, each unfolding will double the number of holes visible. Unfolding the piece with one hole once produces a piece with 2 holes. Unfolding a second time produces a piece with 4 holes. Unfolding the third time produces a piece with 8 holes. Unfolding fourth time produces a piece with 16 holes. Finally, unfolding for the fifth time piece produces the original sheet of paper with 32 holes.

13. (D) 6

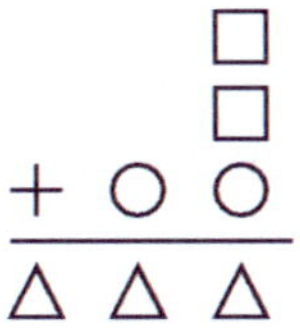

The sum is less than $10 + 10 + 100 = 120$, so each triangle represents 1 and the sum is 111. If the circle represents 8 (or less), then the sum does not exceed $9 + 9 + 88 = 106$, and then the sum could not be 111. Therefore, each circle represents 9. $111 - 99 = 12$, so each square represents $12 \div 2 = 6$. The digits 1, 9, and 6 are different, and the answer is 6. Indeed, $6 + 6 + 99 = 12 + 99 = 111$.

14. (B) 108 g

255 g is the weight of 3 apples and 2 oranges and 285 g is the weight of 2 apples and 3 oranges, so (255 + 285) g is the weight of (3 + 2) apples and (2 + 3) oranges. Thus, 540 g is the weight of 5 apples and 5 oranges. The combined weight of 1 apple and 1 orange is $540\text{ g} \div 5 = 108$ g.

15. (B) 2

Thomas' statement, "This number is equal to 9," contradicts both Andrew and Michael since 9 is neither even nor 15, so Roman's statement, "This number is prime," must be true. Since 15 is not prime, Michael's statement is false, so Andrew's statement, "This number is even," must be true. The only number that is both prime and even is 2, so the boys are talking about the number 2.

16. (B) 2

A square has 4 axes of symmetry as shown below, one horizontal, one vertical, and two diagonal. For each line of symmetry, the small squares that need to be shaded are shown in green. For the horizontal line there are 3 green squares, for the vertical line there are 5 green squares, for the first diagonal line there are 3 green squares, and for the other diagonal line there are 2 green squares.

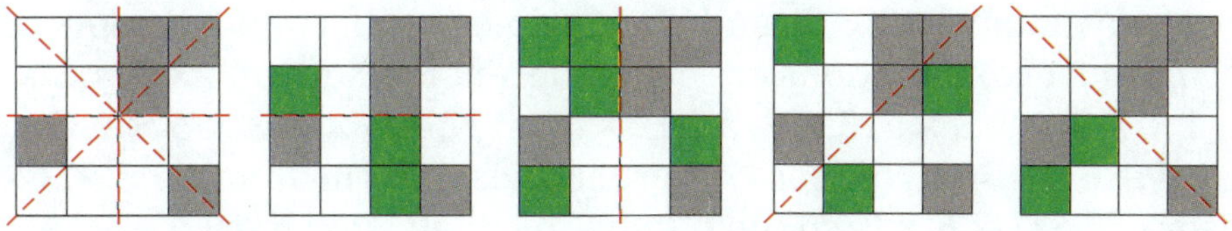

Thus, 2 is the smallest number of little squares that have to be shaded in order to get a line of symmetry for the figure.

17. (E) 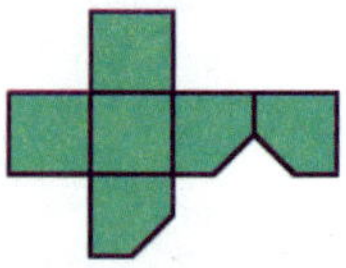

The cube shows 3 faces without one vertex each. Each of (A), (B), and (C) shows only 2 faces without a vertex, so none of them represents the plan of the cube after unfolding. After folding plan (D), the missing vertices do not come together at one corner, so (D) is not a valid plan for the cube. In (E), after folding the plan, the missing vertices do come together at one corner.

18. (C) 35 dm

Snail Fin moves 25 dm along 5 diagonal segments, so one diagonal segment has the length $d = 25 \text{ dm} \div 5 = 5$ dm. Snail Pin moves 37 dm along 5 diagonal segments and 4 vertical segments, so the total length of 4 vertical segments is 37 dm – 25 dm = 12 dm. One vertical segment has the length $v = 12 \text{ dm} \div 4 = 3$ dm. Snail Rin moves 38 dm along 6 vertical segments and 5 horizontal segments, so the total length of the 5 horizontal segments is 38 dm – (6 × 3 dm) = 20 dm. One horizontal segment has the length $h = 20 \text{ dm} \div 5 = 4$ dm. Snail Tin moves along 3 diagonal segments, 4 vertical segments, and 2 horizontal segments. The total length of these segments is $3d + 4v + 2h = 3 \times 5 \text{ dm} + 4 \times 3 \text{ dm} + 2 \times 4 \text{ dm} = 35$ dm. Hence, snail Tin has gone 35 dm.

19. (C) Thursday

During any 7 consecutive days there are exactly 4 sunny days. During the last 6 weeks of the vacation, from the 3rd day to the 44th day there are 6 × 7 days and 6 × 4 of them are sunny. During the first two days of the vacation we can have two consecutive sunny days if they are Thursday and Friday, so Thursday should be the first day of vacation to enjoy as many sunny days as possible.

20. (D) 44

If f is the first number and s is the second number, then f + s = 77 and 8 × f = 6 × s. One way to solve the equations is to multiply the first equation by 8, so we can replace 8 × f with 6 × s. 8 × 77 = 8 × (f + s) = 8 × f + 8 × s = 6 × s + 8 × s = 14 × s = 2 × 7 × s, so 2 × 7 × s = 8 × 77 and this simplifies to s = 4 × 11 = 44. f = (6 × 44) ÷ 8 = 33 (or f = 77 – 44 = 33). Both f and s are natural numbers and 44 is the larger of the two numbers.

21. (C) 16

1 is a divisor of any number. 2, 3, 5, and 7 are prime numbers and divisors of 2 × 3 × 5 × 7, so the following product combinations are also different divisors of 2 × 3 × 5 × 7:
2 × 3, 2 × 5, 2 × 7, 3 × 5, 3 × 7, 5 × 7; and 2 × 3 × 5, 2 × 3 × 7, 2 × 5 × 7, 3 × 5 × 7.
There are 1 + 4 + 6 + 4 = 15 of these different divisors. The number 2 × 3 × 5 × 7 is itself the 16th divisor of 2 × 3 × 5 × 7, so the number of all divisors is 16.

22. (B) 36

The fraction $\frac{5}{9}$ can't be simplified, so the number of Ella's mushrooms is a multiple of 9. The fraction $\frac{2}{17}$ can't be simplified, so the number of Ola's mushrooms is a multiple of 17. 70 is neither a multiple of 9 nor a multiple of 17, so the number of Ola's mushrooms is 17, 34, 51, or 68. The corresponding number of Ella's mushrooms is $70 - 17 = 53$, $70 - 34 = 36$, $70 - 51 = 19$, or $70 - 68 = 2$. Only 36 is a multiple of 9, so Ella has 36 mushrooms.

23. (B) 8

Any 4 consecutive cells contain two groups of 3 consecutive cells with the two middle cells shared by both groups. The sum of numbers in any 3 consecutive cells is the same, so the numbers in the first and the fourth cell of any 4 consecutive cells must be the same. Hence, every third cell contains the same number. Thus, the numbers in the 1st, 4th, 7th, and 10th cell are the same; the numbers in the 2nd, 5th, 8th, and 11th cell are the same; and the numbers in the 3rd, 6th, and 9th cell are the same. 6 is in the 9th cell, so it is also in the 3rd cell. 7 is in the 1st cell and the sum of numbers in the first three cells is 21. Therefore, $21 - 7 - 6 = 8$ must be placed in the 2nd cell. Here is the complete solution:

7	8	6	7	8	6	7	8	6	7	8

24. (A) $\frac{1}{4}$

The picture shows 8 triangles with one common vertex. Along each edge of the big square there are two triangles (one shaded and the other not shaded) with a common height and the ratio of their bases 4 to 1. Hence, the area of the triangle that is not shaded is $\frac{1}{4}$ of the area of the shaded triangle. This is true for all four edges, so the area of the figure that is not shaded is $\frac{1}{4}$ of the area of the shaded figure.

25. (C) \$28.50

Let p be the initial price of each of the two CDs. The price of the first CD goes down to $0.95 \times p$ and the price of the second CD goes up to $1.15 \times p$. The difference of \$6.00 between the new prices is equal to $(1.15 \times p - 0.95 \times p)$, so $0.2 \times p = \$6.00$.
$p = \$30.00$ and the cheaper CD is now $0.95 \times \$30.00 = \$28.50$.

26. (A) 128

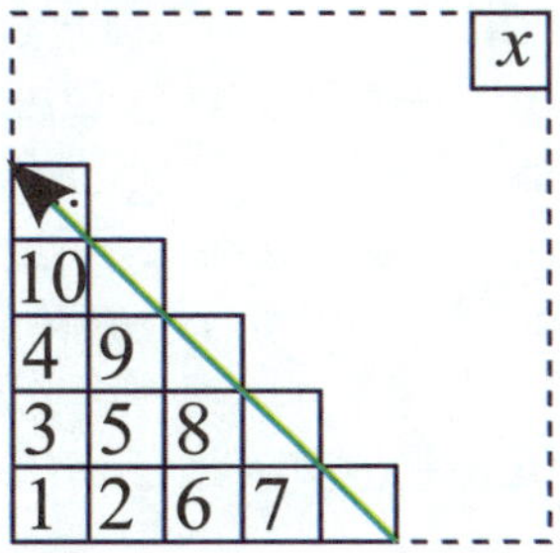

The numbers are written successively in the direction parallel to the diagonal connecting the upper left corner with the lower right corner of the big square. We start at the small square with 1 and end at the small square with the letter x, so the largest number of the table is written in the cell with x. The table with n rows and n columns contains n^2 cells, so the largest number in the table is n^2 when we start with 1, write consecutive numbers, and never repeat any numbers. Thus, the number in the cell with x is n^2; in other words, it's a perfect square. 128 is not a perfect square, so it cannot be placed in the cell with x. The other options ($256 = 16^2$, $81 = 9^2$, $121 = 11^2$, and $400 = 20^2$) are perfect squares.

27. (D) 667

$2004 = 668 \times 3$, so $\underbrace{111111\ldots111}_{2004\ times}$ can be grouped as $\underbrace{111,111,\ldots,111}_{668\ times}$.

$111 \div 3 = 37$; $111,111 \div 3 = 37,037$; $111,111,111 \div 3 = 37,037,037$; and so on.

Hence, $\underbrace{111,111,\ldots,111}_{668\ times} \div 3 = 37\,\underbrace{037,037,\ldots,037}_{667 times}$.

Thus, there are 667 zeros in the result of dividing $\underbrace{111111\ldots111}_{2004\ times}$ by 3.

28. (D) 8

The number of marbles in one box must be a divisor of 108 and 180, so it is a common divisor of 108 and 180. The number of boxes is the smallest when the common divisor is the greatest. $108 = 3 \times 36$ and $180 = 5 \times 36$, so 36 is the greatest common divisor of 108 and 180. You need $108 \div 36 = 3$ boxes for the red marbles and $180 \div 36 = 5$ boxes for the green marbles.

29. (A) 17

If a student gives S correct answers out of 10 questions, then they earn $5 \times S - 3 \times (10 - S) =$ $= (8 \times S - 30)$ points. If the student earned E points, then $E = (8 \times S - 30)$ and $S = (30 + E) \div 8$.

Hence, Mathew gave $(30 + 34) \div 8 = 64 \div 8 = 8$ correct answers;
Philip gave $(30 + 10) \div 8 = 40 \div 8 = 5$ correct answers;
and John gave $(30 + 2) \div 8 = 32 \div 8 = 4$ correct answers.

Altogether they answered $(8 + 5 + 4) = 17$ problems correctly.

30. (C) 18 cm^2

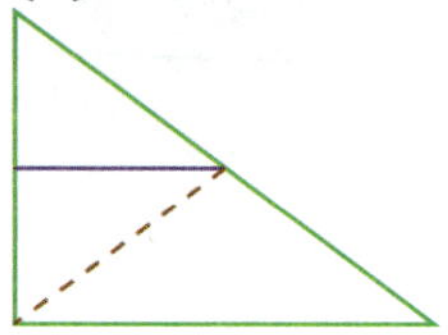

The area of the right triangle is $(6\text{ cm} \times 8\text{ cm}) \div 2 = 24\text{ cm}^2$.
Among the options given as possible answers we can show that 18 cm^2 as the area of the resulting polygon can happen. A particular folding line (in blue) is parallel to one leg and passes through the midpoint of the other leg. The area of the triangle above the folding line is 6 cm^2, so the area of the trapezoid below the folding line is 18 cm^2.

For any triangle and any folding line dividing the triangle into two pieces, the resulting polygon has an area strictly smaller than the given triangle since there is always some overlap along the folding line. If a folding line divides a triangle into two pieces with different areas, then the resulting polygon has an area strictly greater than half of the area of the given triangle since folding can only increase (or keep the same) the area of each of the two pieces.

If a folding line divides a given triangle into two pieces with the same area but the folding line is not an axis of symmetry for the given triangle, then folding increases the area of at least one piece, so the resulting polygon has an area greater than half of the area of the given triangle.

If the folding line is an axis of symmetry of the given triangle, then both pieces are congruent triangles, so the folding line passes though one vertex of the given triangle and the midpoint of the opposite side. This happens only when the given triangle is isosceles. Our right triangle is not isosceles, so the area of the resulting polygon is never 12 cm^2.

In a summary, the area of the resulting polygon must be greater than 12 cm^2 and less than 24 cm^2. The only option in this range is 18 cm^2.

Solutions for Year 2006

1. (D) 2006
$2005 + 2007 + a = 2006 - 1 + 2006 + 1 + a = 2 \times 2006 + a$. If $3 \times 2006 = 2005 + 2007 + a$, then $2 \times 2006 + a = 3 \times 2006$, so $a = 2006$.

2. (E) 7,645,413,092
7 is the greatest first digit among cards shown, so it goes first. After that, each time we have to pick a card from the cards not yet used, we pick the card with the highest first digit.

3. (D) 22

There are 10 places along the top row, 10 places along the bottom row, 1 place at the left side, and 1 place at the right side. Altogether, there are 22 places at the rectangular table.

4. (B) $60.00

 and  together cost $90.00,

so 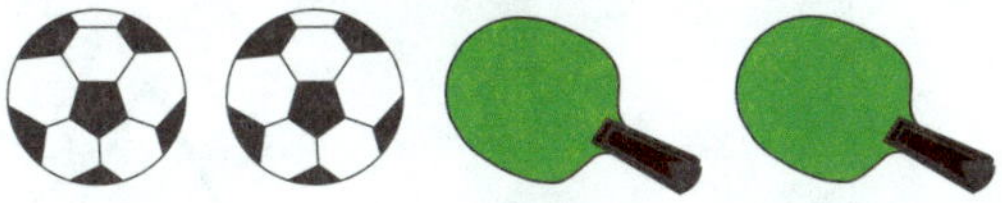 cost $180.00.

Since  + 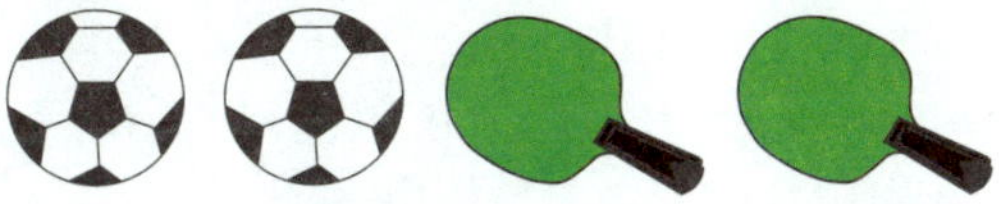 cost $240,

one soccer ball costs $240.00 – $180.00 = $60.00.

5. (E) 

The twelve numbers divide the clock face into 12 equal arcs. Each arc represents 30° since $360 \div 12 = 30$. Hence, 150° will be made up of 5 such arcs, because $150° \div 30° = 5$. This is represented by the angle between 12 and 5.

6. (A) 37

When counting the consecutive odd numbers, starting with 1, use the formula $\frac{1+Odd}{2}$; when counting the consecutive even numbers, starting with 2, use the formula $\frac{Even}{2}$.

$\frac{1+39}{2} = 20$, so 39 is the 20th odd number. $\frac{34}{2} = 17$, so 34 is the 17th even number.
There are 37 houses on Long Street since $20 + 17 = 37$.

7. (D) 8

Starting from the 2, we can move either to the right (R) or the left (L) to pick up the next number. All paths make 2006. There are 8 paths: 2LLL, 2LLR, 2LRL, 2LRR, 2RLL, 2RLR, 2RRL, and 2RRR.

8. (A) 0.005

One hundredth in decimal notation is 0.01 and half of it is $0.01 \div 2 = 0.005$.

9. (C) 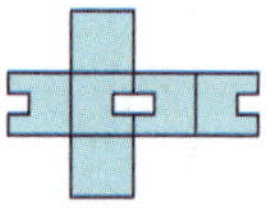

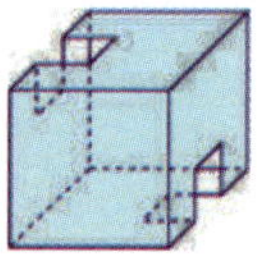

The openings are at two parallel edges that are not on the same face. This excludes (A) and (D). Each opening is in the middle of its edge, which excludes (E). (B) has only one full opening and two half-openings that do not form one full opening in the cube. (C) is a perfect match for the box.

10. (D) 91

$(5 + 6)^2 - (5 \times 6) = 121 - 30 = 91$

11. (D) 160

The diameter of one circle is $2 \times 5 = 10$. The side length of the square is equal to two diameters, so it is 20. Hence, all sides of the equilateral triangles are equal to 20. 8 of the triangle sides form the perimeter of the four-pointed star, so the perimeter is equal to $8 \times 20 = 160$.

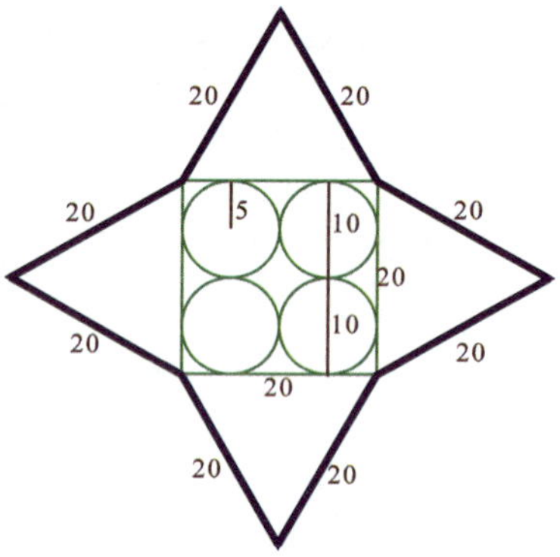

12. (D) 1000

2, 4, 6, …, 1996, 1998, 2000 are the first 1000 consecutive positive even numbers and 1, 3, 5, …, 1995, 1997, 1999 are the first 1000 consecutive positive odd numbers.
The difference between the sum of 1000 even numbers and the sum of 1000 odd numbers is the same as the sum of the following 1000 differences:
$(2 - 1), (4 - 3), (6 - 5), \ldots, (1996 - 1995), (1998 - 1997), (2000 - 1999)$.
Each of the differences is 1 and the sum of 1000 of these differences is 1000.

13. (E) equilateral triangle

If P and Q are points on one piece of paper and, after folding the paper, P and Q become one point, then the folding line is the bisector of the segment PQ.

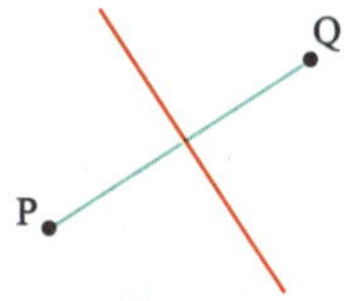

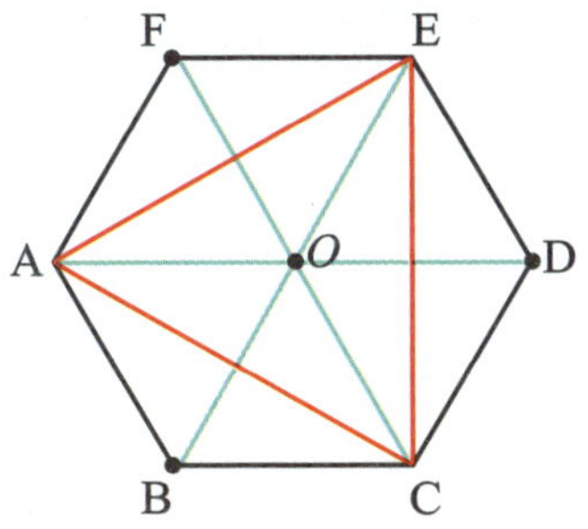

ABCDEF is a regular hexagon and O is its center.
OA = AB = BO = OC = CB since ΔBOA and ΔBOC are equilateral triangles. OABC is a rhombus and the diagonals of any rhombus are bisectors of each other. Thus, the lines AC, CE, and EA are the folding lines for OB, OD, and OF, respectively.
Once the hexagon is folded this way along the three lines it is entirely inside the triangle ACE, which is an equilateral triangle.

14. (A) 2

One side of a 3 × 3 × 3 cube consists of 9 squares. The cube has 6 sides, so its surface consists of 6 × 9 = 54 squares.
The white region of the solid shown consists of 12 squares.
To paint 54 squares, 9 fluid ounces of paint are needed, so
$\frac{12}{54} \times 9$ fluid ounces $= \frac{12}{6}$ fluid ounces = 2 fluid ounces of paint are needed to paint the white region.

15. (B) 90

1 hour = 60 minutes = 60 × 60 seconds.
25 meters per second is 25 × 60 × 60 = 90,000 meters per hour.
Since 1 kilometer is 1,000 meters, the car will travel 90 kilometers in one hour.

16. (A) $\frac{1}{4}$

F is the midpoint of AE and E is the midpoint of AB, so $|FE| = ½|AE| = ½(½|AB|) = 1$ inch since $|AB| = 4$ inches. H is the midpoint of AG and G is the midpoint of AD, so
$|HG| = ½|AG| = ½(½|AD|) = \frac{1}{4}$ inch since $|AD| = 1$ inch.
The area of the shaded rectangle (in square inches) is $\frac{1}{4} \times 1 = \frac{1}{4}$.

17. (B) 1010101010

First perform the five subtractions (the results are shown below) and then add these 5 results.

$$\begin{array}{rr} & 1\,1\,1\,1\,1\,1\,1\,1\,1\,1 \\ - & 1\,1\,1\,1\,1\,1\,1\,1\,1 \\ + & 1\,1\,1\,1\,1\,1\,1\,1 \\ - & 1\,1\,1\,1\,1\,1\,1 \\ + & 1\,1\,1\,1\,1\,1 \\ - & 1\,1\,1\,1\,1 \\ + & 1\,1\,1\,1 \\ - & 1\,1\,1 \\ + & 1\,1 \\ - & 1 \\ \hline \end{array}$$

$$\begin{array}{rr} & 1\,0\,0\,0\,0\,0\,0\,0\,0\,0 \\ + & 1\,0\,0\,0\,0\,0\,0\,0 \\ + & 1\,0\,0\,0\,0\,0 \\ + & 1\,0\,0\,0 \\ + & 1\,0 \\ \hline & 1\,0\,1\,0\,1\,0\,1\,0\,1\,0 \end{array}$$

18. (C) 20

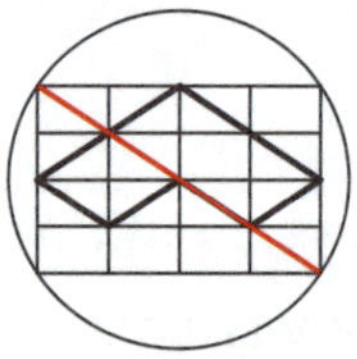

The diameter of the circle (shown in red) is 10 and it consists of 4 diagonals of small rectangles. The perimeter of the figure marked with the bold line consists of 8 diagonals of small rectangles. The length of 4 diagonals is 10, so the length of 8 diagonals is 20

19. (B) 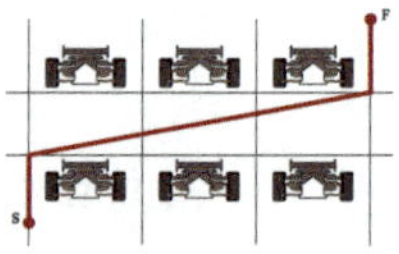

Under each picture the corresponding path is shown schematically.

(A) 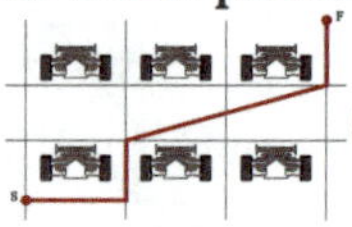 (B) 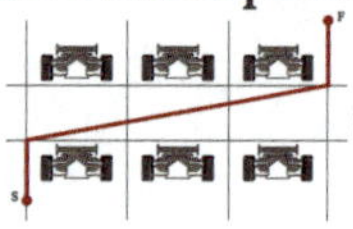 (C) 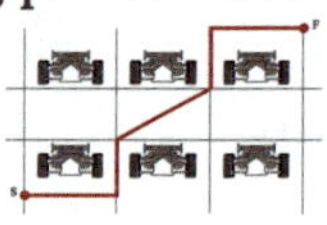 (D) 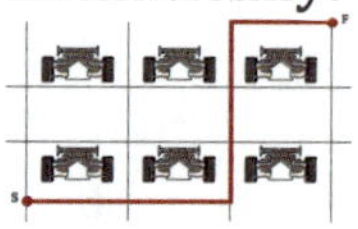

(A') 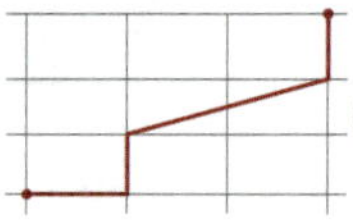 (B') 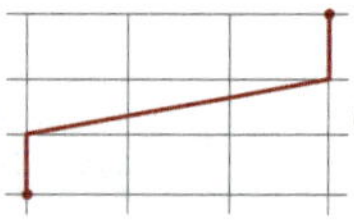 (C') 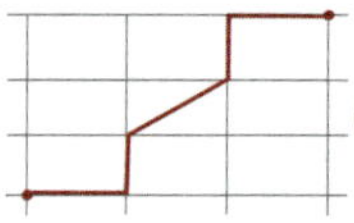 (D') 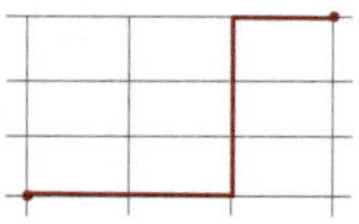

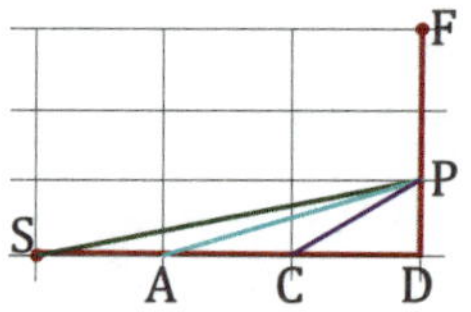

In each case we reroute the path without changing its length. We move horizontally first (there is no horizontal segment for path (B)), then we move along the slanted segment (there is no slanted segment for path (D)), and finally we move up vertically two segments (three vertical segments for path (D)). All rerouted paths pass through point P and the segment PF is common for all four rerouted paths. Notice that

$|SP| + |PF|$ = the length of path (B),

$|SA| + |AP| + |PF|$ = the length of path (A),

$|SC| + |CP| + |PF|$ = the length of path (C), and

$|SD| + |DP| + |PF|$ = the length of path (D).

To compare the lengths, we only need the Triangle Inequality: in any triangle the sum of the lengths of two sides is greater than the length of the third side.

Consequently, $|SP| < |SA| + |AP|$, $|AP| < |AC| + |CP|$, and $|CP| < |CD| + |DP|$.

Therefore, the length of the path (B) < the length of the path (A) < the length of the path (C) and the length of the path (C) < the length of the path (D).

Thus, the path (B) is the shortest.

20. (B) 3

The biggest two-digit number divisible by 3 is 99 and the smallest two-digit number divisible by 3 is 12, so Anne's sum is $99 + 12 = 111$.

The biggest two-digit number not divisible by 3 is 98 and the smallest two-digit number not divisible by 3 is 10, so Adam's sum is $98 + 10 = 108$.

The difference between their sums is $111 - 108 = 3$.

21. (E) *B, A, C*

$|OB| = |OE| - |BE| = 2006 - 1111 = 895$, $|OA| = 1111$, and $|OC| = 0.70 \times |OE| = 1404.2$, so from O to E the points go in the order *B, A, C*.

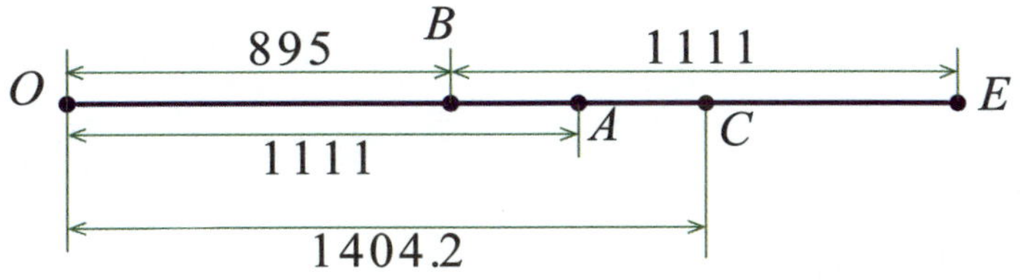

22. (B) 4

$1 + 2 + 3 + 4 + 5 = 15$ is the smallest sum of different natural numbers that does not exceed 15. The rope was divided into 5 pieces, so 4 cuts were made.

23. (D) 6

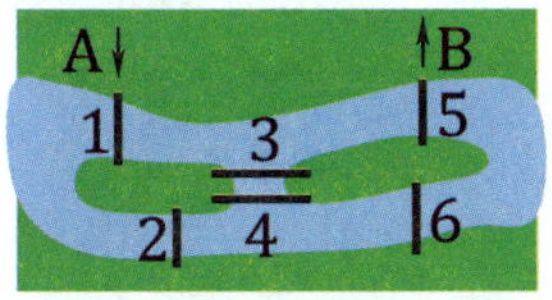

The possible routes are: 1 2 6 3 4 5; 1 2 6 4 3 5; 1 3 4 2 6 5; 1 3 6 2 4 5; 1 4 3 2 6 5; and 1 4 6 2 3 5.

Each time you have to pick a bridge, pick the one with the smallest number that you have not used yet.

24. (C) $\frac{1}{10}, \frac{9}{80}, \frac{1}{8}$

On a number line, the midpoint of a segment is the average of the endpoints, which is a half of the sum of the endpoints. The largest and smallest numbers are the endpoints, so we need to check if their average is equal to the middle number.

In (A), the smallest number is $\frac{1}{5}$ and the largest number is $\frac{1}{3}$, but

$$\left(\frac{1}{5} + \frac{1}{3}\right) \div 2 = \left(\frac{3}{15} + \frac{5}{15}\right) \div 2 = \frac{8}{15} \times \frac{1}{2} = \frac{4}{15}, \text{ and not } \frac{1}{4};$$

in (B), $(12 + 32) \div 2 = 44 \div 2 = 22$, and not 21;

in (C), $\left(\frac{1}{10} + \frac{1}{8}\right) \div 2 = \left(\frac{4}{40} + \frac{5}{40}\right) \div 2 = \frac{9}{40} \times \frac{1}{2} = \frac{9}{80}$;

in (D), $(0.3 + 1.3) \div 2 = 1.6 \div 2 = 0.8$, and not 0.7;

in (E), $(24 + 64) \div 2 = 88 \div 2 = 44$, and not 48.

The only set where the midpoint is listed along with endpoints is answer (C), as $\frac{9}{80}$ is the midpoint of the segment with the endpoints $\frac{1}{10}$ and $\frac{1}{8}$.

25. (E) 124

is a 1 × 1 square. It has (**1** + 1) rows of sticks and (**1** + 1) columns of sticks. Each of the rows and columns consists of **1** stick, so there are (**1** + 1) × **1** + (**1** + 1) × **1** = 2 × [(**1** + 1) × **1**] sticks.

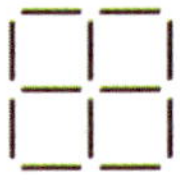

is a 2 × 2 square. It has (**2** + 1) rows of sticks and (**2** + 1) columns of sticks. Each of the rows and columns consists of **2** sticks, so there are (**2** + 1) × **2** + (**2** + 1) × **2** = 2 × [(**2** + 1) × **2**] sticks.

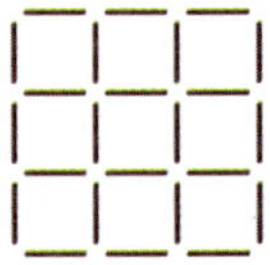

is a 3 × 3 square. It has (**3** + 1) rows of sticks and (**3** + 1) columns of sticks. Each of the rows and columns consists of **3** sticks, so there are (**3** + 1) × **3** + (**3** + 1) × **3** = 2 × [(**3** + 1) × **3**] sticks.

The 4th square is a 4 × 4 square. It has (**4** + 1) rows of sticks and (**4** + 1) columns of sticks. Each of the rows and columns consists of **4** sticks, so there are (**4** + 1) × **4** + (**4** + 1) × **4** = = 2 × [(**4** + 1) × **4**] sticks. Following the pattern, we can state that the 31st square consists of 2 × [(**31** + 1) × **31**] sticks and the 30th square consists of 2 × [(**30** + 1) × **30**] sticks.
The difference between them is 2 × 32 × 31 – 2 × 31 × 30 = 2 × 31 × (32 – 30) = 2 × 31 × 2 = = 4 × **31** = 124 sticks, so Barbara will use 124 more sticks to create the 31st square from the 30th square.

Here is another solution:

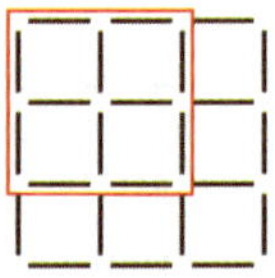

If you remove the 2 × 2 upper left corner (as shown in red) from the 3 × 3 square, then you are left with the bottom row and the rightmost column of the 3 × 3 square. Be aware that they share one square at the lower right corner.
For each square in the bottom row (including the corner) you count 2 sticks in the form  and for each square in the rightmost column (including the corner) you count 2 sticks in the form . In this way you count all the sticks that are needed to create the **3** × **3** square when adding to the 2 × 2 square. The number of new sticks is **3** × 2 + **3** × 2 =12. By the same argument, once Barbara created the 30th square, she needs **31** × 2 + **31** × 2 = 4 × **31** = 124 more sticks to create the 31st square.

26. (D) E

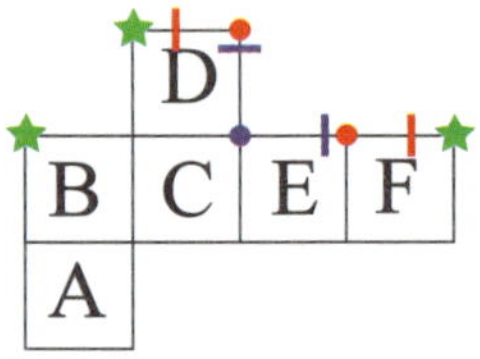

After folding the diagram:

(1) the blue point becomes the vertex D-C-E , which is the common vertex of faces D, C, and E;

(2) the red points become one point—the vertex D-E-F, which is the common vertex of faces D, E, and F;

(3) the green stars become one point—the vertex D-F-B, which is the common vertex of faces D, F, and B;

(4) the edges marked by blue marks become one edge—the D-E edge, which connects the blue and red vertices of the cube; the letter D faces the E face, so E will be placed next to it in the second net;

(5) the edges marked by red marks become one edge—the D-F edge, which connects the red and green vertices of the cube.

27. (A) 21

We can convert the percentages to decimals to make the calculations more straightforward. At the first station 0.30 of the gas was taken out, so 0.70 remained in the tank. At the second station 0.40 of the 0.70 of the tank, which is $0.40 \times 0.70 = 0.28$, was taken out. After the two stops $(0.30 + 0.28)$ of the gas was taken out, so $1 - (0.30 + 0.28) = 0.42$ of the gas was still in the tank. At the third station ½ of the 0.42, which is 0.21, was taken out. Altogether, at the three stations $0.30 + 0.28 + 0.21 = 0.79$ of the gas was taken out, so $1 - 0.79 = 0.21 = 21\%$ of the initial amount of gas was left in the tank.

28. (B) 1

$\frac{1}{8} + \frac{1}{6} + \frac{2}{3} = \frac{3+4+16}{24} = \frac{23}{24}$, so $1 - \frac{23}{24} = \frac{1}{24}$ of all the students in the class received an F.
The number of students in the class is less than 30 and it must also be a multiple of 24. Therefore, the class has 24 students and 1 student received an F.

29. (E) \$18 for Adam and \$12 for Tom

Paul's \$30 was the price for 15 tickets, so \$2 was the price for one pool ticket per person. Adam bought $8 \times 3 = 24$ tickets and paid $24 \times \$2 = \$48$ for them. Tom bought $7 \times 3 = 21$ tickets and paid $21 \times \$2 = \$42$ for them. Since each boy's tickets cost \$30, Adam should get $\$48 - \$30 = \$18$ and Tom should get $\$42 - \$30 = \$12$ of Paul's \$30.

30. (C) 501

12 is the least common multiple of 2, 3, and 4. It is the first number underlined three times.
From 1 to 12 there are 3 numbers underlined exactly two times; they are 4, 6, and 8.
From 13 to 24 there are 3 numbers underlined exactly two times; they are 16, 18, and 20.
Notice that $13 = 1 + 12$, $24 = 12 + 12$, $16 = 4 + 12$, $18 = 6 + 12$, and $20 = 8 + 12$.
From 25 to 36 there are 3 numbers underlined exactly two times; they are 28, 30, and 32.
Notice that $25 = 13 + 12$, $36 = 24 + 12$, $28 = 16 + 12$, $30 = 20 + 12$, and $32 = 20 + 12$.
The pattern continues, so from 37 to 48 there are 3 numbers underlined exactly two times.
The same is true for intervals from 49 to 60, from 61 to 72, from 73 to 84, and so on.
2004 is a multiple of 3 and 4, so it is a multiple of 12. Hence, from 1993 to 2004 there are 3 numbers underlined exactly two times.
Count the number of intervals ending with consecutive multiples of 12 and multiply the final count by 3. $2004 = 167 \times 12$, so there are 167 such intervals and $167 \times 3 = 501$.
2005 is not underlined and 2006 is underlined just once, so there are 501 numbers from 1 to 2006 underlined exactly twice.

Solutions for Year 2008

1. (C) $2 \times 0 \times 0 \times 8$
In (A), $2 + 0 + 0 + 8 = 10$.
In (B), $200 \div 8 = 25$.
In (C), $2 \times 0 \times 0 \times 8 = 0$.
In (D), $200 - 8 = 192$.
In (E), $8 + 0 + 0 - 2 = 6$.
The smallest result is 0, so (C) $2 \times 0 \times 0 \times 8$ is the smallest number.

2. (C) 2×3
$2 \times 2 \times 3 \times 3 = (2 \times 3) \times (2 \times 3)$, so we have to replace  with 2×3 to make the expression true.

3. (B) PJN
Starting with 3:
the result of JPN is $[(3 \times 3) + 2] - 1 = 10$;
the result of PJN is $[(3 + 2) \times 3] - 1 = 14$;
the result of JNP is $[(3 \times 3) - 1] + 2 = 10$;
the result of NJP is $[(3 - 1) \times 3] + 2 = 8$; and
the result of PNJ is $[(3 + 2) - 1] \times 3 = 12$. Only the order PJN ends up with 14.

4. (D) 0
Let's substitute the symbol with the options given in each answer and check for which one the result of the left-hand side equals 100.
(A) $1 + 1 + 1 - 2 = 1$
(B) $1 + 1 - 1 - 2 = -1$
(C) $1 + 1 \div 1 - 2 = 1 + 1 - 2 = 0$
(D) $1 + 101 - 2 = 100$, and
(E) $1 + 111 - 2 = 110$.
Hence, to make the expression $1 + 1 \clubsuit 1 - 2 = 100$ true, we need to replace $\clubsuit$ with 0.

5. (E) 

When you overlap two triangles then at least one of the six vertices of these two triangles is also a vertex of the resulting figure. We have two non-right triangles, so the resulting figure has at least one angle not equal to 90°, so we cannot get a rectangle as the resulting figure. The diagrams below show we can get the other four figures by overlapping the two given identical triangles.

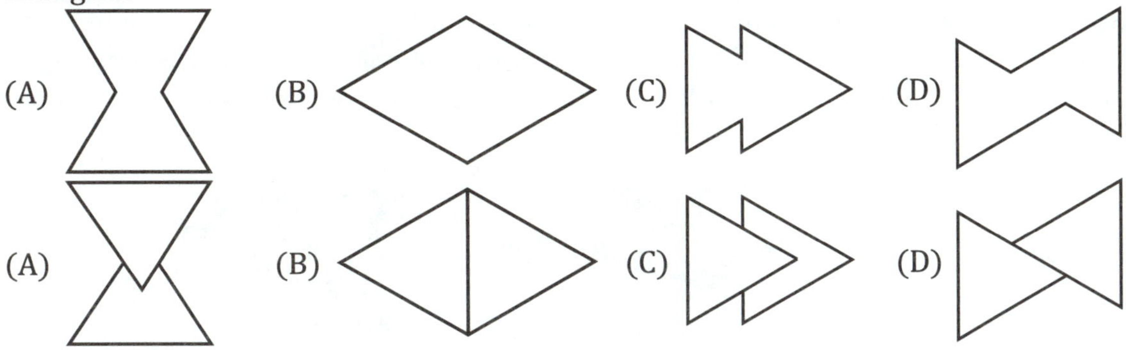

6. (C) 2

For each flag shown below we compute the ratio of the black area to the total area.

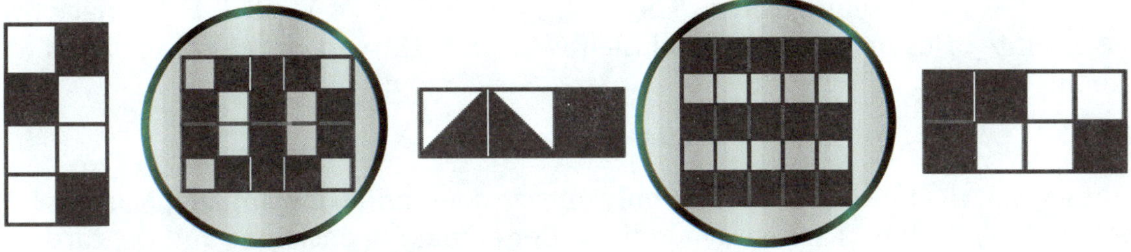

For the 1st flag the ratio is $\frac{3}{8}$. For the 2nd flag the ratio is $\frac{12}{20} = \frac{3}{5}$.

For the 3rd flag the ratio is $\frac{2}{3}$ since there are 3 squares, with one square completely black and 2 halves (an equivalent of 1 square) also black.

For the 4th flag the ratio is $\frac{15}{25} = \frac{3}{5}$ and for the 5th flag the ratio is $\frac{4}{8} = \frac{1}{2}$, so only for the 2nd and the 4th flag the ratio is $\frac{3}{5}$. Therefore, 2 flags were made correctly.

7. (B) 6

The sum of the numbers in the first row is 9 and the sum of the numbers in the second row is 6, so the sum of all four numbers is $9 + 6 = 15$. 2, 3, and 4 are the three given numbers, so the fourth number is $15 - (2 + 3 + 4) = 6$.

To get the sum of 9 in the first row we have to write 6 and 3 (in any order) in the first row and then write 2 and 4 (in any order) in the second row.

3	6
2	4

8. (E) 19
$17 + Paul's own money was enough to buy a $21 present and have $15 left over, so after borrowing the money Paul had $36. Since he had borrowed $17, he had $36 – $17 = $19 in his piggy bank.

9. (A) 54
Let's call the unknown multipliers: a, b, c, and d.

×	a	b
c	35	63
d	30	?

Numbers 35 and 30 are both multiples of 5. This gives us a = 5. (1 is also a factor of both 35 and 30, but the second number in the first row is not a multiple of 63.) Then, c = 7 and d = 6. Since c = 7, b = 9 since the product must be 63. The question mark would represent b times d, which is $9 \times 6 = 54$.

10. (E) 14

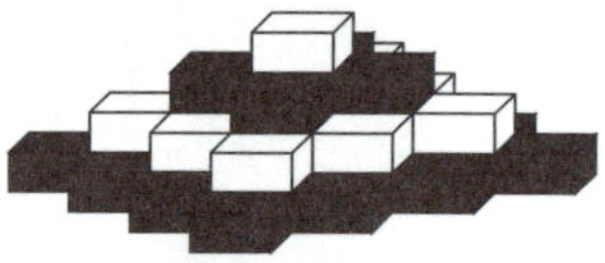
Figure 1

Figure 2

Each layer of the tower is made of bricks of the same color. The bottom layer consists of $1 + 3 + 5 + 7 + 5 + 3 + 1 = 25$ black bricks. The layer above it consists of $1 + 3 + 5 + 3 + 1 = 13$ white bricks. The next layer consists of $1 + 3 + 1 = 5$ black bricks. There is 1 white brick on the very top of the tower. $1 + 13 = 14$, so 14 white bricks were used to make this tower.

11. (E) O
We want each of the five boxes to have only one card left and we want all five cards left to be different. Below, the boxes are shown in the order of making decisions and the cards which are removed are marked in red.

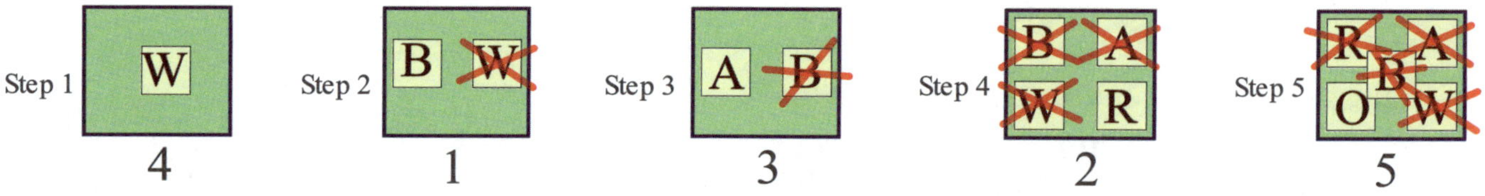

At each step there is only one option to keep a card that is different from the cards remaining from the previous steps. The card left in box 5 is the card with the letter O.

12. (B) 24 cm
The perimeter of the square is 4×4 cm = 16 cm. The triangle and the square share one side, so one side of the triangle is 4 cm and the sum of the lengths of the other two sides of the triangle is 16 cm – 4 cm = 12 cm. These two sides of the triangle and three sides of the square form the perimeter of the pentagon. Hence, the perimeter of the pentagon is 12 cm + 3×4 cm = 24 cm.

13. (D) 4

All the matches are identical, so we use 1 match as a unit of measurement for the sides of triangles. With 4 matches we can get 3 separate segments only if their lengths are 2, 1, and 1. In any triangle the sum of lengths of any two sides is always greater than the length of the third side, but $1 + 1$ is not greater than 2, so we cannot build a triangle using 4 identical matches. In each of the other four cases we can built a triangle. Here are the lengths of sides for the triangles made using the other numbers of matches: $7 = 3 + 3 + 1$ or $7 = 3 + 2 + 2$; $6 = 2 + 2 + 2$; $5 = 2 + 2 + 1$; and $3 = 1 + 1 + 1$.

14. (E) 256 cm^2

A square with an area of 121 cm^2 is an 11 cm × 11 cm square. The bigger square has the dimensions $(11 + 5)$ cm × $(11 + 5)$ cm, so its area is 16 cm × 16 cm = 256 cm^2.

15. (D) $\frac{1}{6}$

$\frac{2}{3}$ of the juice was left in the bottle after $\frac{1}{3}$ was poured into a glass. $\frac{3}{4} \times \frac{2}{3} = \frac{2}{4} = \frac{1}{2}$ of juice was poured into a pitcher, so $\frac{1}{3} + \frac{1}{2} = \frac{5}{6}$ of the juice was poured out of the bottle. Therefore, $\frac{1}{6}$ of the original amount of juice was left in the bottle.

16. (D) 9

+	0	2	3	6
0	0	2	3	6
2		4	5	8
3			6	9
6				12

The table to the left represents all possible scores when throwing two darts. The different scores are: 0, 2, 3, 4, 5, 6, 8, 9, and 12. There are 9 of them.

17. (C) 69

Alexa has $3 \times 7 = 21$ CDs in 3 boxes and another 2 CDs on her desk. Thus, Alexa has 23 CDs not on the shelf. This is one third of all her CDs, so the number of all her CDs is $3 \times 23 = 69$.

18. (C) 

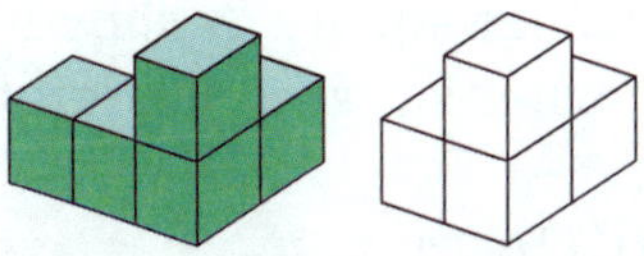

The original figure and all the other figures except (C) contain the core section made of 4 blocks shown as the white figure farther to the right. The one block of the original figure that is not a part of the core section can be easily moved to transform the original figure into any of the figures below except (C). For figure (C), we have to move two blocks.

19. (B) 901

$10^1 - 1 = 9$, $10^2 - 1 = 99$, $10^3 - 1 = 999$, $10^4 - 1 = 9999$, $10^5 - 1 = 99999$, … and so on.

Hence, $10^{101} - 1 = \underbrace{999 \ldots 999}_{101\ times}$ and $10^{101} - 9 = (10^{101} - 1) - 8 = \underbrace{999 \ldots 999}_{101\ times} - 8 = \underbrace{999 \ldots 999}_{100\ times}1$.

The sum of the digits is $100 \times 9 + 1 = 901$.

20. (C) The son and the daughter are the same age.
In two years, the son will be 4 years older than he was two years ago. If he is going to be twice the age he was then, he was 4 years old two years ago, will be 8 years old in two years, and now is 8 – 2 = 6 years old.
In three years, the daughter will be 6 years older than she was three years ago. If she is going to be three times as old as she was then, she was 3 years old 3 years ago, will be 9 in three years, and is 9 – 3 = 6 years old now.
Therefore, the son and the daughter are the same age, and all the other statements are false.

21. (E) 9
@ can't be 0 since @ + @ + @ = * and @ and * have to be different digits. # can't be 0 since # + # + # = & and # and & have to be different digits. @ and # are not zeros and are different digits, so (@ + #) is at least 3. ^ = * + & = (@ + @ + @) + (# + # + #) = 3 × (@ + #).
^ must be 9 since ^ is a digit and the smallest value for 3 × (@ + #) is 3 × 3 = 9.

22. (C) Roberts, Smith, Farrell
The doctor is the youngest of the three friends, so Farrell is not the doctor since Farrell is older than the engineer. Also, Farrell is not the engineer. Hence, Farrell must be the musician.
The doctor does not have a sister, so Smith is not the doctor since Smith has a sister (married to the musician Farrell).
Hence, Smith must be the engineer. Consequently, Roberts is the doctor.
Therefore, Roberts is the doctor, Smith is the engineer, and Farrell is the musician.

23. (D) on any of the shaded squares
The robot can move only from a non-shaded square to a shaded square or from a shaded square to a non-shaded square, so a path through all 9 squares could be described either as SNSNSNSNS or NSNSNSNSN, where S stand for "shaded" and N stands for "non-shaded." There are only 4 non-shaded squares, which excludes any NSNSNSNSN path because it requires five Ns. So, the robot cannot start on any non-shaded square.

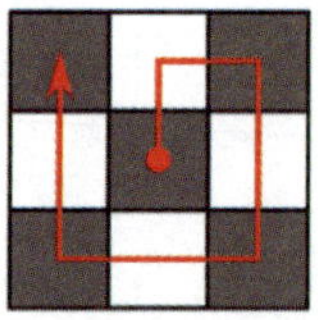

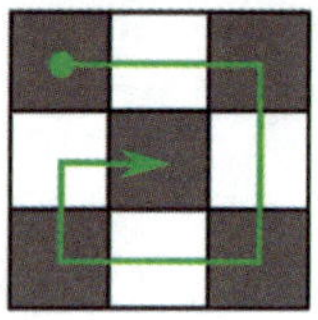

It can start on the middle square as shown by the red path and it can start on a corner square as shown by the green path. By symmetry, it can also start on any corner square. Therefore, the robot can start on any of the shaded squares.

24. (C) 9 km
Notice that the lengths of route L1 and L2 added together are equal to the lengths of routs L3 and L4 added together. 17 km + 12 km = 29 km.
29 km – 20 km = 9 km.

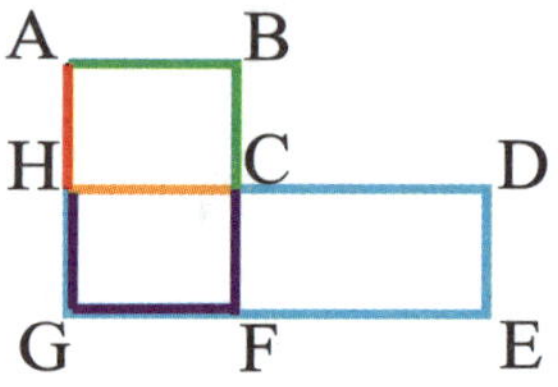

L_1, the length of the route L1, is CD + DE + EF + FG + GH + HC;
L_2, the length of the route L2, is AB + BC + CF + FG + GH + HA;
L_3, the length of the route L3, is AB + BC + CD + DE + EF + FG + GH + HA;
L_4, the length of the route L4, is CF + FG + GH + HC, so
$L_1 + L_2$ = CD + DE + EF + FG + GH + HC + AB + BC + CF + FG + GH + HA =
= [AB + BC + CD + DE + EF + FG + GH + HA] + [CF + FG + GH + HC] = $L_3 + L_4$, so
$L_4 = L_1 + L_2 - L_3 = 17 + 12 - 20 = 9$. Thus, the route of bus L4 is 9 kilometers long.

25. (D) 25

We want to select a pair of points so that the other two points are between the two points selected. There are 6 possible selections for the two points farthest from each other.

(1) Points A and B: this cannot be the case because $|CD| > |AB|$, so the points C and D cannot both be inside segment AB.

(2) Points B and C: this cannot be the case as CD is longer than BC, so point D cannot be inside segment BC.

(3) Points C and D: A is between C and D because $|AC| = |CD| - |DA| =$ $= 14 - 12 = 2$. However, B cannot be between A and D since $|AB|$ is 13 and $|DA|$ is only 12. So, C and D is not the pair of points farthest from each other.

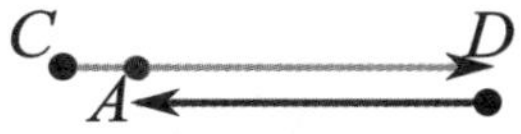

(4) Points D and A: this cannot happen because AB is longer than DA, so point B cannot be inside segment DA.

(5) Points A and C: If point D is between A and C, then $|CA| = |CD| + |DA| = 14 + 12 = 26$. If point B is also between A and C, then $|AC| = |AB| + |BC| = 13 + 11 = 24$. $|CA| = |AC|$ but $26 \neq 24$, so this cannot actually happen.

(6) Points B and D: This is actually the case. $|DA| + |AB| = 12 + 13 = 25$ and $|BC| + |CD| =$ $= 11 + 14 = 25$, so the distance between the two points farthest from each other is 25.

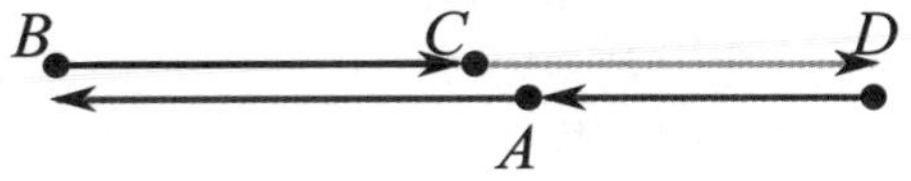

The correct arrangement of all the points is shown above.

26. (C) 50 meters

It takes 12 seconds from the time the front of the train enters the bridge until the rear of the train enters it. Since it takes 60 seconds from the time the front of the train enters the bridge until the rear of the train exits the bridge, it will take $60 - 12 = 48$ seconds for the rear of the train to pass through a whole bridge that is 200 meters long. Since 12 seconds is one quarter of 48 seconds, the length of the train is one quarter of the length of the bridge. One quarter of 200 meters is 50 meters, so the train is 50 meters long.

27. (D) 32 oz

The combined volume of the 6 bottles is $(16 + 18 + 22 + 24 + 32 + 34)$ oz, which is 146 oz. When the empty bottle is removed, the other five bottles must be split into two groups so that the ratio of the volumes is 2 : 1, since there is twice as much orange juice as cherry juice.
Hence, (146 oz – the volume of the empty bottle) must be a multiple of 3.
$146 - 16 = 130$ is not a multiple of 3, $146 - 18 = 128$ is not a multiple of 3,
$146 - 22 = 124$ is not a multiple of 3, $146 - 24 = 122$ is not a multiple of 3,
$146 -$ **32** $= 114$ **is a multiple of 3**, and $146 - 34 = 112$ is not a multiple of 3.
Therefore, the volume of the empty bottle is 32 oz.
Since ⅓ of 114 is 38, the bottles with cherry juice are the 16 and 22 oz bottles $(16 + 22 = 38)$.
Since ⅔ of 114 is 76, the bottles with orange juice are the 18, 24, and 34 oz bottles $(18 + 24 + 34 = 76)$.

28. (D) 45
The differences for numbers from 10 to 19 are: 1, 0, -1, -2, -3, -4, -5, -6, -7, -8 and the differences for numbers from 80 to 89 are: 8, 7, 6, 5, 4, 3, 2, 1, 0, -1.
The sum of these 20 differences is 0. The same is true for numbers from 20 to 29 and 70 to 79; for numbers from 30 to 39 and 60 to 69; and numbers from 40 to 49 and 50 to 59.
The differences for numbers from 90 to 99 are: 9, 8, 7, 6, 5, 4, 3, 2, 1, 0. Their sum is $(9+0)+(8+1)+(7+2)+(6+3)+(5+4)=5\times 9=45$.
Therefore, the sum of all of the results for numbers from 10 to 99 is 45.

29. (D) 5
$\frac{n+41}{n+5}=\frac{(n+5)+36}{n+5}=\frac{n+5}{n+5}+\frac{36}{n+5}=1+\frac{36}{n+5}$, so $\frac{n+41}{n+5}$ is a natural number if $n+5$ is a divisor of 36.
Remember that n is a positive integer, so $n+5$ is at least 6 and it is a divisor of 36.
The options are $n+5=6, 9, 12, 18$, or 36. Hence, $n=1, 4, 7, 13$, or 31, and there are 5 natural numbers for which the quotient $\frac{n+41}{n+5}$ is a natural number.

30. (C) 746
The 1000-digit number 20082008…2008 contains 500 zeros, 250 twos, and 250 eights.
We want to erase as many small digits as possible, so we should start by finding the largest digits we need to keep. The sum of 250 eights is $250\times 8=2000$. We only need 4 more twos to get the sum of 2008. Hence, we can erase all 500 zeros and 246 twos, and the sum of the remaining digits is 2008. Thus, $500+246=746$ is the largest number of digits that can be erased so that the sum of the remaining digits is 2008.

Solutions for Year 2010

1. (B) 3
If we remove two ▲ from each side, the equation ▲ + ▲ + 6 = ▲ + ▲ + ▲ + ▲ simplifies to 6 = ▲ + ▲, so ▲ represents 3.

2. (D) $2+0+1+0$
$201+0=201=3\times 67$, so it is not prime.
$2+0-1+0=1$, which is not prime (1 is neither prime nor composite).
20×10 is not prime as it is a multiple of two integers.
$2+0+1+0=3$, which is prime.
$20\times 1\times 0=0$, which is not prime.
Thus, only $2+0+1+0$ is prime.

3. (C) 11

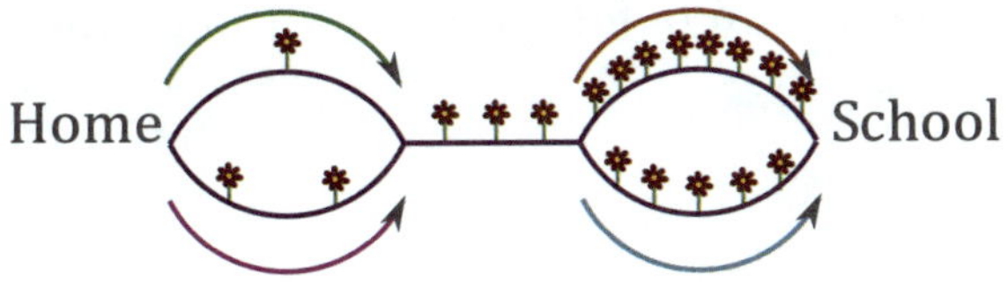

Ella can walk from home to school along 4 different paths. Each path is identified by the number of flowers she passes.
Path 1: $1 + 3 + 8$; Path 2: $1 + 3 + 5$;
Path 3: $2 + 3 + 8$; and Path 4: $2 + 3 + 5$.
The sums are 12, 9, 13, and 10. Among the five answer options only 11 is not one of them.

4. (C) 30

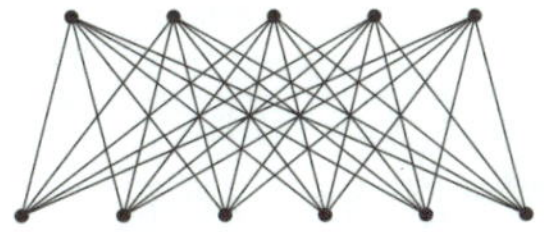

There are 5 points at the top and 6 points at the bottom. Each point at the top is connected to each point at the bottom, so there are $5 \times 6 = 30$ segments. Thus, Anna drew 30 segments.

5. (C) in 3 years
Mr. Meow is 13 years old, so two years ago he was 11 years old. Mr. Whiskers was 4 years old two years ago since $11 + 4 = 15$. Now Mr. Whiskers is 6 years old since $4 + 2 = 6$. He will be 9 years old in 3 years since $6 + 3 = 9$.

6. (B) 2, 4, 6, and 8
The cuts are made at every other numbered edge. Rotate Figure 1 so it matches the orientation of Figure 2. You can see that the cuts are along edges with even labels. You could rotate Figure 1 further by 90° each time to see other settings matching Figure 2. Each time the cuts are along edges with even numbers.

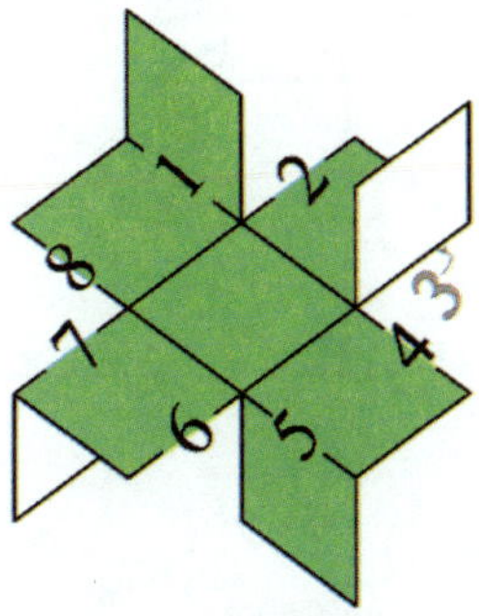

7. (B) 2
By moving just one bar we cannot make room for a new bar. Thus, we have to move at least 2 bars. One way this can be done is shown to the right. We move the top horizontal bar to the left and then slide the vertical bar up into the free space. This creates a space for a new bar.

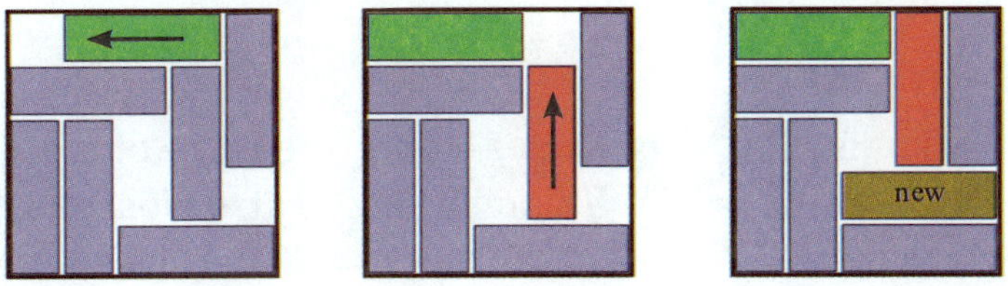

8. (B) 23
The missing page numbers are 29, 30, 31... , 72, 73, and 74. There are $74 - 28 = 46$ missing page numbers. Each sheet contains two page numbers, so the number of missing sheets is half of 46, which is 23.

9. (E) $20 \div 10 \times 20 + 10$

Follow the order of operations: do multiplication and division before doing addition and subtraction.

(A) $20 \times 10 + 20 \times 10 = 200 + 200 = 400$,
(B) $20 \div 10 \times 20 \times 10 = 2 \times 20 \times 10 = 400$,
(C) $20 \times 10 \times 20 \div 10 = 200 \times 20 \div 10 = 4000 \div 10 = 400$,
(D) $20 \times 10 + 10 \times 20 = 200 + 200 = 400$, and
(E) $20 \div 10 \times 20 + 10 = 2 \times 20 + 10 = 40 + 10 = 50$.

Thus, $20 \div 10 \times 20 + 10$ has a different value than the other expressions.

10. (C) 4 cats

2 flies and 3 spiders have $2 \times 6 + 3 \times 8 = 36$ legs. 10 birds have $10 \times 2 = 20$ legs and the difference between 36 legs and 20 legs is 16 legs. One cat has 4 legs, so $16 \div 4 = 4$ cats have 16 legs. Thus, 2 flies and 3 spiders have as many legs as 10 birds and 4 cats.

11. (E) $6 \times 5 + 8 \times 2$

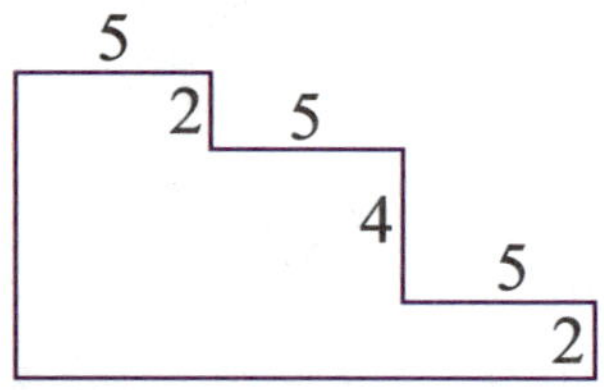

The lengths of the vertical segments of the stairs add up to $2 + 4 + 2$, which is 4×2. This is also the length of the left edge of the figure. The lengths of the horizontal segments of the stairs add up to $5 + 5 + 5$, which is 3×5. This is also the length of the bottom edge of the figure. Hence, the perimeter of the figure is $2 \times (3 \times 5 + 4 \times 2) = 6 \times 5 + 8 \times 2$.

12. (E) 728

Reverse the steps in the calculation. $777 \div 7 = 111$, $111 - 7 = 104$, and $104 \times 7 = 728$, so Adam's secret number was 728.

13. (D) 135°

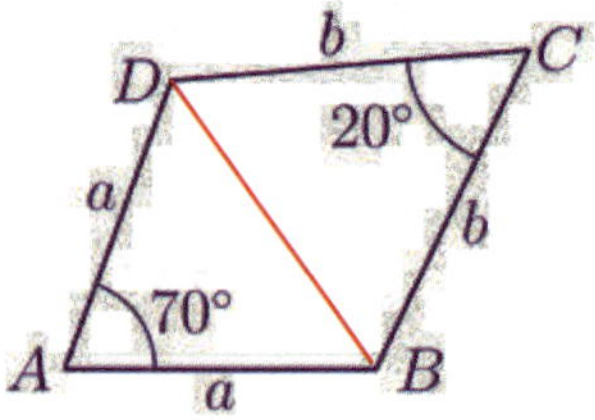

Triangle BAD is isosceles, so the measure of $\angle ABD$ is $(180° - 70°) \div 2 =$ $= 110° \div 2 = 55°$. Triangle BCD is also isosceles, so the measure of $\angle DBC$ is $(180° - 20°) \div 2 = 160° \div 2 = 80°$. $\angle ABD + \angle DBC = \angle ABC$, so angle ABC measures $55° + 80° = 135°$.

14. (D) 800 cm²

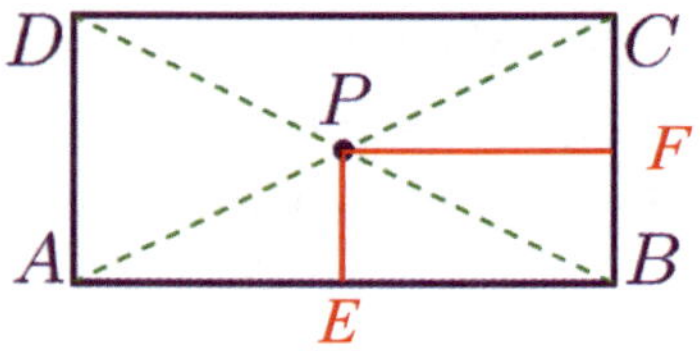

$ABCD$ is a rectangle with a perimeter of 120 cm. E is the midpoint of AB, F is the midpoint of BC, and P is the center of the rectangle. Since the length of PF is twice the length of PE, the length of AB is twice the length of BC, so the sum of the lengths AB and BC is 3 times the length of BC. This sum is also half of the perimeter of the rectangle $ABCD$, so $3 \times$ the length of $BC = 60$ cm. Thus, the length of BC is 20 cm and the length of AB is 40 cm. Therefore, the area of the rectangle is $40 \text{ cm} \times 20 \text{ cm} = 800 \text{ cm}^2$.

15. (D) 8

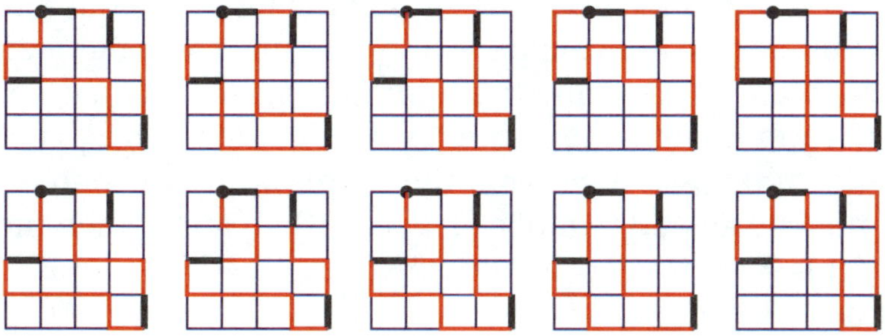

Any reasonable boundary line must be within the 4 × 4 square that contains the four bold segments.
The ant has many options to choose a path that encloses 8 unit squares but there is no path that encloses 7 or fewer unit squares.

16. (C) 

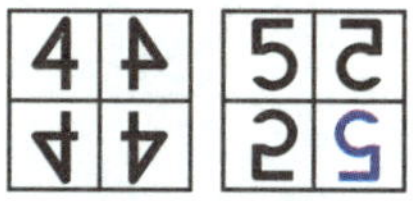

Figure 1 shows that folding the 2 × 2 square, unfolding it and then folding it again produces images such that any two images in adjacent cells are symmetric with respect to the common edge of the cells. In effect, the image in the lower right-hand corner is upside down and flipped. If **4** in the upper left corner is replaced by **5**, then the complete figure is shown to the right. It matches option (C).

17. (A) $\frac{1}{3}$

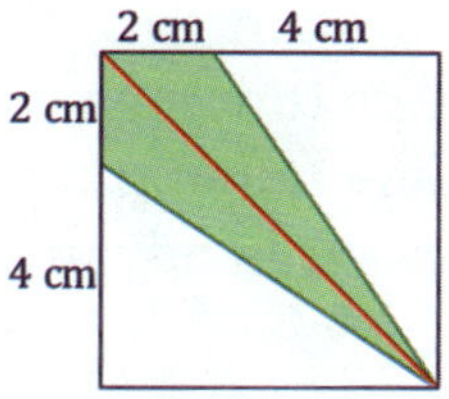

The shaded region and the square are symmetric with respect to the diagonal (shown in red). Below the diagonal, the shaded triangle and the unshaded triangle have a common height (the bottom edge) and the base of the unshaded triangle is twice the base of the shaded triangle. Because the area of a triangle is ½ × base × height, the area of the unshaded triangle is twice the area of the shaded triangle. By symmetry, the unshaded area of the square is twice the shaded area of the square, so the area of the whole square is three times the shaded area. Therefore, the shaded region is $\frac{1}{3}$ of the square.

18. (C) 6

There are 5 rows, so we can only have 5 black squares. There are 11 black squares now, so we have to paint 6 of the black squares white. We must do it in such a way that there is also only one black square in every column. It can be done as shown to the right. The numbers indicate the black squares painted white.

19. (C) 14

For the middle die, the sum of the dots on the two faces glued to the other two dice is 7 since the sum of the number of dots on opposite faces of every die is always 7. For the right die, there are 3 dots on the face glued to the middle die since the opposite face has 4 dots. The left die is identical to the right die and is placed in exactly the same way, so the face glued to the middle die has 4 dots. Hence, the sum of the number of dots on the faces that have been glued together is $4 + 7 + 3 = 14$.

20. (E) 24

$1 + 4 + 7 + 10 + 13 = 35$ and 1, 4, 7, 10, and 13 are in the cells of the figure, so the sum of numbers in the row + the sum of numbers in the column = 35 + the unknown number at the center (it is counted twice). We want the row and the column sums to be equal, so 35 + the unknown number at the center is the same as the result of adding two equal numbers, which is always even. Thus, 35 + the unknown number at the center is an even number. Hence, the number at the center must be an odd number (1, 7, or 13). If it is 1, then the row sum must be $(35 + 1) \div 2 = 18$; if it is 7, then the row sum must be $(35 + 7) \div 2 = 21$; if it is 13, then the row sum must be $(35 + 13) \div 2 = 24$.

There is an actual solution for each of the three options. Possible solutions are shown below. The largest of these sums is 24.

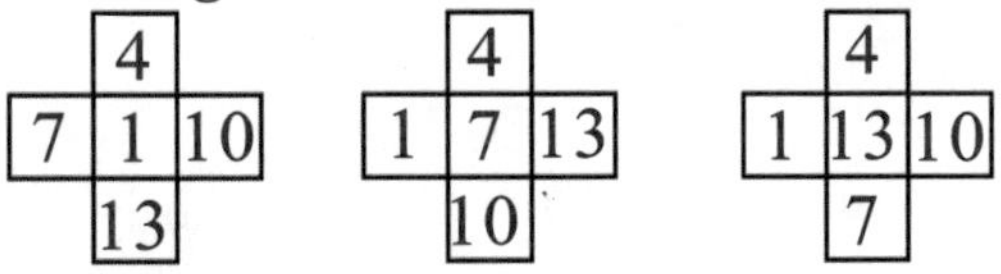

21. (D) 45

There are 15 different types of topping combinations (listed below) and 3 different sizes of pizza, so there are $3 \times 15 = 45$ possible different pizzas.

All possible combinations of toppings are:

all four toppings	(1) p, h, m, o					
three toppings	(2) p, h, m	(3) p, h, o	(4) p, m, o	(5) h, m, o		
two toppings	(6) p, h	(7) p, m	(8) p, o	(9) h, m	(10) h, o	(11) m, o
one topping	(12) p	(13) h	(14) m	(15) o		

Note: In advanced mathematics, the answer can be written as $3 \times (2^4 - 1)$.

22. (C) 22 mm

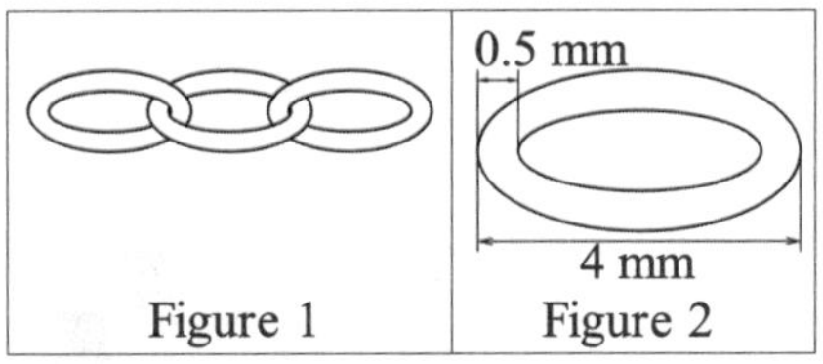

Figure 1 Figure 2

The two-link overlap has the length of 2×0.5 mm = 1 mm. The first link has the length of 4 mm and each additional link will only increase the length of the chain by 4 mm – 1 mm, which is 3 mm.

A chain made of seven links will be 22 mm long since 4 mm + 6×3 mm = 4 mm + 18 mm = 22 mm.

Another calculation, 0.5 mm + 7×3 mm + 0.5 mm = 22 mm, is based on the picture below.

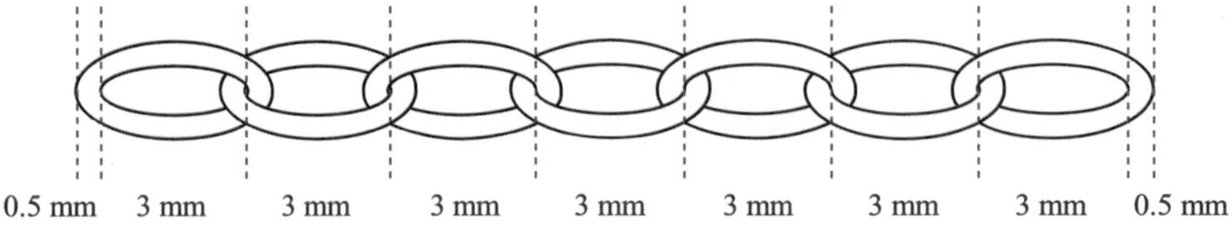

23. (B) 52

Counting backwards from Adam's 4th chair, which is Tom's 20th chair, we can see that Tom's 17th chair is Adam's 1st chair.

Adam's chair	4th	3rd	2nd	1st
Tom's chair	20th	19th	18th	17th

Adam's 46th chair is Tom's 10th chair. Tom counts the next 6 chairs, his 11th, 12th, 13th, 14th, 15th, and 16th chair, before reaching his 17th chair, which is the same as Adam's 1st chair. So, after Adam's 46th chair there are 6 more chairs to count before Adam reaches his 1st chair. $46 + 6 = 52$, so there are 52 the chairs at the table.

24. (B) 1056

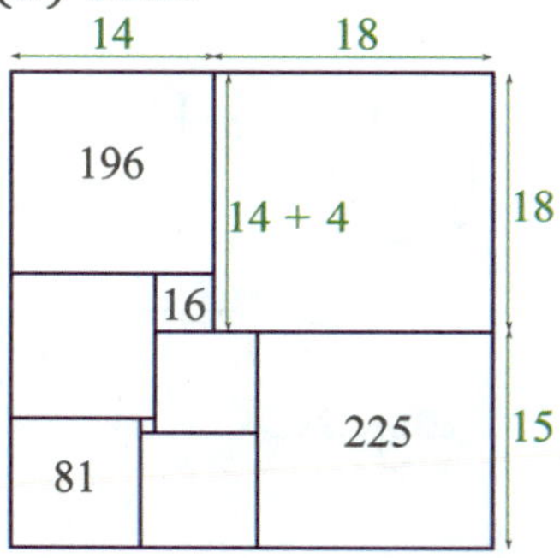

The square with the area equal to 196 has a side length equal to 14 since $14 \times 14 = 196$. The square with the area equal to 16 has a side length equal to 4 since $4 \times 4 = 16$. So, the square at the upper right corner has a side length equal to $14 + 4 = 18$. The top side of the rectangle has the length 32 since $14 + 18 = 32$. Any square with the area 225 has a side length equal to 15 ($15 \times 15 = 225$), so the right side of the rectangle has the length $18 + 15 = 33$. Therefore, the area of the rectangle is $32 \times 33 = 1056$.

The figure to the right shows the areas of all the squares.

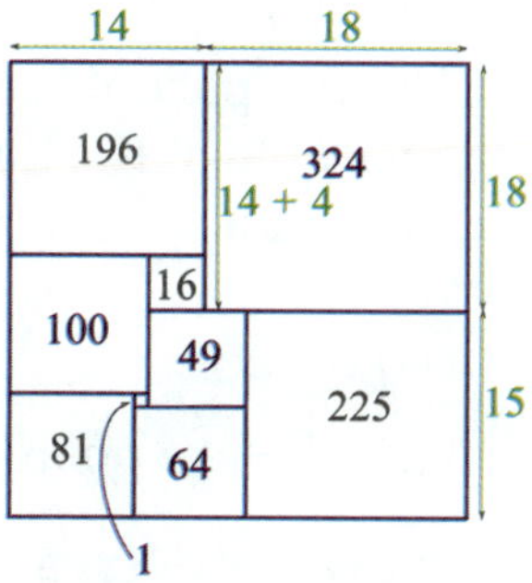

25. (E) 96

5 is a prime number, so the product of any number of digits is equal to 5 only if one digit is 5 and all other digits are equal to 1. The sum $5 + \underbrace{1 + 1 + 1 + \cdots + 1 + 1 + 1}_{n\ times} = 100$ only for $n = 95$, so we are counting 96-digit numbers with one digit that is 5 and all other digits that are 1. The digit 5 can be in any of the 96 positions, so there are 96 numbers that have the sum of their digits equal to 100 and the product of the digits equal to 5.

26. (B) 6

We are looking at numbers $1t82u$ such that the digits $1, t, 8, 2, u$ are different and the number $1t82u$ is a multiple of 4 and 3 (so it is a multiple of 12). The number $1t82u$ is a multiple of 4 only if $2u$, which equals $20 + u$, is a multiple of 4. It happens for $u = 0, 4,$ and 8. However, 8 cannot be used for u since all the digits must be different.

For $u = 0$, the numbers are in the form $1t820$. The number $1t820$ is a multiple of 3 only if the sum $1 + t + 8 + 2 + 0 = 11 + t$ is a multiple of 3. It happens for $t = 1, 4,$ and 7. However, 1 cannot be used for t since all the digits must be different.

So far only 14820 and 17820 are numbers with the required properties.

For $u = 4$, the numbers are in the form $1t824$. The number $1t824$ is a multiple of 3 only if the sum $1 + t + 8 + 2 + 4 = 15 + t$ is a multiple of 3. It happens for $t = 0, 3, 6,$ and 9, which gives us the numbers 10824, 13824, 16824, and 19824.

Altogether, there are 6 numbers with the required properties.

27. (E) 118
The right side of the theater has seats with even numbers. The numbers in the first row are 2, 4, 6, 8, 10, 12, 14, 16, 18, and 20. Then the number 22 starts the next row and this row stretches to the number 40. The pattern is shown in the table.
Anna's seat number was 100. Seat number 99 is on the other side of the aisle, so it is not as close as the other seats. Among seats numbered 76, 94, 104, and 118, seat number 118 is closest to Anna's seat, so Eve's ticket was for seat number 118.

102	**104**	106	108	110	112	114	116	**118**	120
82	84	86	88	90	92	**94**	96	98	100
62	64	66	68	70	72	74	**76**	78	80
42	44	46	48	50	52	54	56	58	60
22	24	26	28	30	32	34	36	38	40
2	4	6	8	10	12	14	16	18	20

28. (B) 

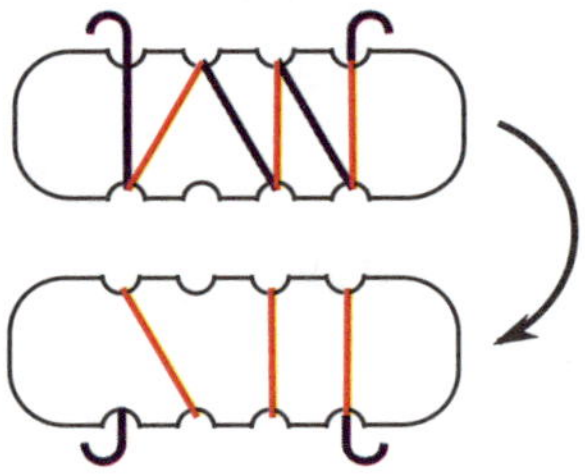

The top part of the picture on the right adds the wrapping on the back side of the board in red. When you flip the board over its long edge, you see the picture (B).

29. (D) 60
m and n are positive integers such that $75 \times m = n^3$. $m + n = \frac{n^3}{75} + n$, so the sum is decreasing when n is decreasing; however $m = \frac{n^3}{75}$ must be a positive integer. n^3 must be a multiple of 75. 75 is a multiple of the prime numbers 3 and 5, so n must be a multiple of 3 and 5. Hence, n must be a multiple of 15. The smallest positive multiple of 15 is 15, so $\frac{15^3}{75} + 15$ is the smallest possible sum if it is a natural number. $\frac{15^3}{75} + 15 = \frac{15\times15\times15}{75} + 15 = \frac{15\times15}{5} + 15 = 3 \times 15 + 15 =$ $= 45 + 15 = 60$, which is a natural number. Therefore, 60 is the smallest possible sum of $m + n$ and it happens when $m = 45$ and $n = 15$.

30. (C) green
Could all four dragons be liars? If so, then they were all seven headed dragons and $4 \times 7 = 28$ would be their number of heads. One of them said just that, so that liar would be telling the truth. Hence, one of the four dragons was a truth-teller. The other three dragons were all liars since two truth-tellers would have to state the same number.
The three liars together had $3 \times 7 = 21$ heads, so all four of them had either $21 + 6 = 27$ heads or $21 + 8 = 29$ heads. No dragon said that altogether they had 29 heads, so that option is not possible. Hence, altogether they had 27 heads and that is what the green dragon stated. Therefore, the dragon who did not lie was green.

Solutions for Year 2012

1. (C) 9
VIVAT KANGAROO contains 9 different letters: V, I, A, T, K, N, G, R, and O. Basil will need 9 different colors.

2. (C) 1.5 m
The whole blackboard is 6 m wide and the middle part is 3 m wide, so together the left-hand and the right-hand parts are 3 m wide. Both side parts have equal width, so the right-hand part of the blackboard is 3 m ÷ 2 = 1.5 m wide.

3. (A) 8
The length of each match corresponds to the diameter of two coins. A square containing 16 coins will have 4 coins in each row and 4 coins in each column. Therefore, each side of the square needs to consist of at least 2 matches. For the whole perimeter, Sally will need $4 \times 2 = 8$ matches.

4. (C) 142
Since there is not row number 13, there are 24 rows on the plane. One of the rows has 4 seats and the other 23 rows have 6 seats each. Thus, the total number of passenger seats is $4 + 23 \times 6 = 4 + 138 = 142$.

5. (E) 6 o'clock this morning
When it is 5 o'clock in the afternoon in Madrid, it is 17 hours after midnight in Madrid. When it is 8 o'clock in the morning in San Francisco, it is 8 hours after midnight in San Francisco. Both of these are on the same day, so Madrid is 17 – 8 = 9 hours ahead of San Francisco.
Ann went to bed in San Francisco at 9 o'clock yesterday evening. Add 9 hours to get the time in Madrid. 3 hours brings us to midnight and the following day, that is, today. We need to add 9 – 3 = 6 more hours, so the time in Madrid is 6 a.m. and it is this morning.

6. (C) 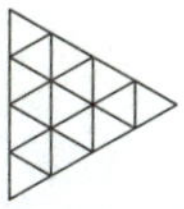

For any 3 neighboring hexagons in the pattern, the centers are vertices of an equilateral triangle with one vertical side, so it is either 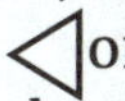 or  .

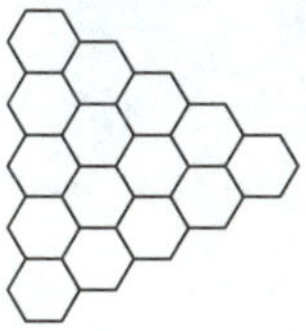

When we move along the outer hexagons and connect the centers of these neighboring hexagons, we form a bigger equilateral triangle. All the small triangles are inside the big equilateral triangle with the vertical side to the left, and there are 16 small triangles inside, so (C) is the answer.

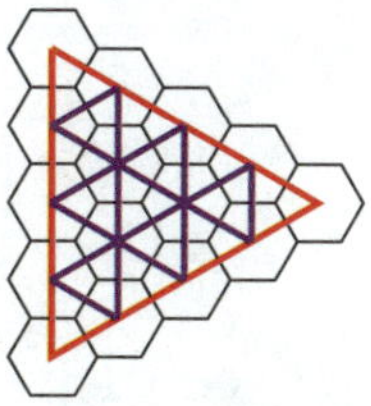

7. (D) $(6 + 3) \times 2 + 1$

First step:	$6 + 3$;
Second step:	the result, $(6 + 3)$, is multiplied by 2, so we get $(6 + 3) \times 2$;
Third step:	add 1 to the second step.

8. (A) 

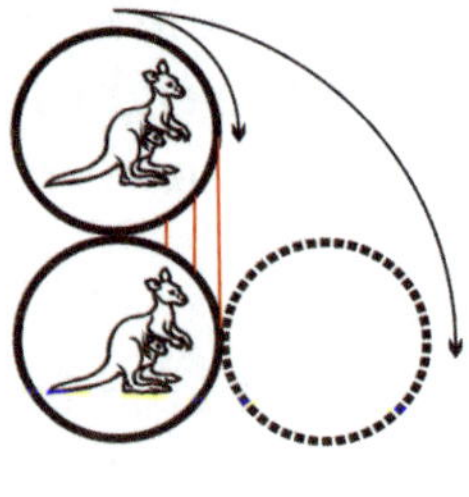

For the rotation, a few matching points of the two coins are connected by vertical red segments. The longest red segment indicates the end of the rotation. At this moment the forearms of the upper coin kangaroo are in front of the forearms of the lower coin kangaroo and the head of the upper coin kangaroo is down. Notice that for each coin, the kangaroo's head is about 60° from its forearms in the counterclockwise direction, so this is true for the upper coin after the rotation. This is exactly the position shown in (A).

9. (B) 20 kg
Adding one balloon allows you to lift an additional 180 kg – 80 kg = 100 kg. Since one balloon can lift items weighing 80 kg plus the basket, the basket must weigh 100 kg - 80 kg = 20 kg.

10. (B) 13
Vivien and Mike ate 9 pieces of fruit (1 apple + 3 pears + 3 apples + 2 pears), so they brought home 25 - 9 = 16 pieces of fruit. One half of the 16 pieces are pears, so they brought home 8 pears. Vivien ate 3 pears and Mike ate 2 pears, so the total number of pears they got from their grandmother was 8 + 3 + 2 = 13.

11. (D) 2, 3, 6
A piece that fits into the center of the puzzle must have a rounded tab and a cutout on opposite sides. Only piece 2 satisfies this condition, so piece 2 goes in the center of the puzzle. The only piece with one straight edge along its right side and tabs on the bottom and on the left is piece 3, so this piece goes next to piece 2. Finally, the only corner pieces are 5 and 6, but only piece 6 has a cutout that will fit above piece 3, so that is the piece that you need to use. The pieces used are 2, 3, and 6.

12. (B) B

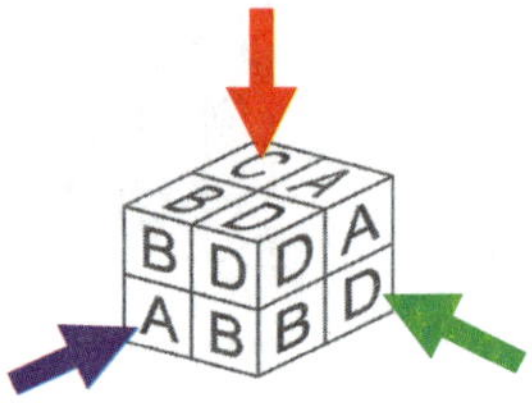

The hidden die is adjacent to the dice marked with arrows in the picture on the left, so it does not have the letter A, C, or D. Therefore, the die has the letter B.

13. (D) 3

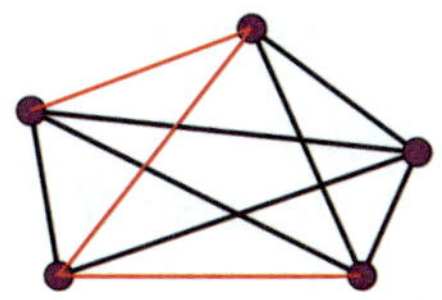

The 3 invisible roads are shown in red. We can count them even without a picture. For any 5 items, call them A, B, C, D, and E, there are always 10 pairs as listed here: {A,B}, {A,C}, {A,D}, {A,E}, {B,C}, {B,D}, {B,E}, {C,D}, {C,E}, and {D,E}. Thus, for 5 cities there are 10 roads connecting all pairs of cities. 7 of them are visible, so 3 roads are invisible.

14. (C) only green

Any number of the form	$1 +$ a multiple of 3	is red,
any number of the form	$2 +$ a multiple of 3	is blue,
and any number of the form	$3 +$ a multiple of 3	is green.

The sum of multiples of 3 is also a multiple of 3, so the sum of a red number and a blue number is ($1 +$ a multiple of 3) + ($2 +$ a multiple of 3) which is $(1 + 2) +$ a multiple of 3.
The resulting sum in the form $3 +$ a multiple of 3 can only be green.

15. (D) 72 cm^2

The perimeter of the figure consists of 14 identical segments, so one segment has the length of $42 \text{ cm} \div 14 = 3 \text{ cm}$. The figure consists of 8 identical squares with the side length of 3 cm, so the area of the figure is $8 \times 3 \text{ cm} \times 3 \text{ cm} = 72 \text{ cm}^2$.

16. (D) 20 cm

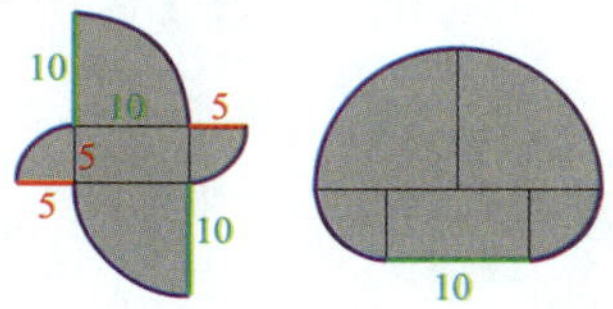

The corresponding curved parts of both perimeters are identical, so the difference between perimeters is exactly the difference between the linear segments of the perimeters, which is
$(5 \text{ cm} + 10 \text{ cm} + 5 \text{ cm} + 10 \text{ cm}) - 10 \text{ cm} = 20 \text{ cm}$.

17. (C) 4

We add 3 numbers along each of the 5 lines. The 5 results are identical, so their sum is a multiple of 5. In this sum of the 5 results, the number at the top is repeated 3 times (it belongs to 3 lines) and any other number belongs to 2 lines and is repeated twice. Hence,
the multiple of 5 $= 2 \times$ **the sum of all seven numbers** $+$ (once more) **the number at the top**.
The sum of the seven numbers is $1 + 2 + 3 + 4 + 5 + 6 + 7 = 28$ and $2 \times 28 = 56$, so
the multiple of 5 $= 56 +$ **the number at the top**. Among the numbers: $56 + 1$, $56 + 2$, $56 + 3$, $56 + 4$, $56 + 5$, $56 + 6$, and $56 + 7$, only $56 + 4 = 60$ is a multiple of 5. Therefore, only 4 can be the number at the top. The top number and the numbers along two horizontal lines add up to 28, so the sum of numbers along one line is $(28 - 4) \div 2 = 12$.

There are twelve solutions for this puzzle. Three of them are shown. The other 9 solutions can be obtained from these solutions by switching the two horizontal lines or the two slanted lines.

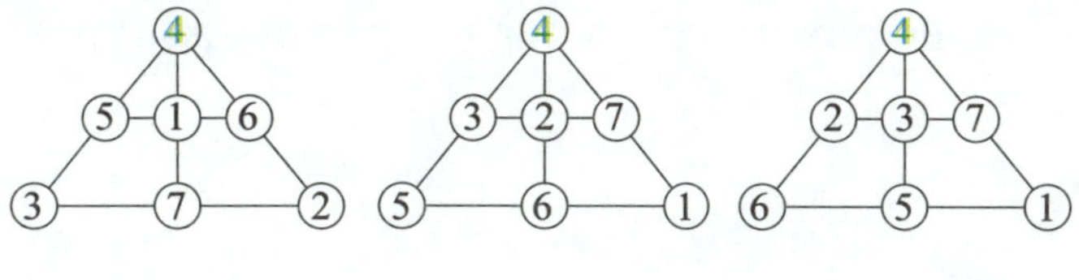

18. (D) 6

$\frac{4}{5} = 0.8$, so after consecutive bounces the ball reaches the heights $10\text{ m} \times 0.8 = 8\text{ m}$, $8\text{ m} \times 0.8 = 6.4\text{ m}$, $6.4\text{ m} \times 0.8 = 5.12\text{ m}$, and $5.12\text{ m} \times 0.8 = 4.096\text{ m}$, respectively.

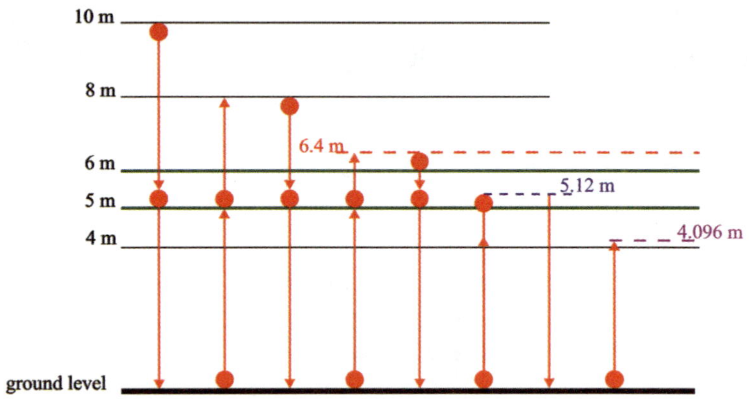

8 phases of the ball's bounces up and down are shown. Each arrow indicates the direction of the motion. The range of the house window is between the two green lines, which are at levels of 5 m and 6 m. Orange balls within this range indicate that the ball appears in front of the window and continues its motion in the current direction of the phase. It happens 6 times. Once the ball reaches the height of 4.096 m it stays below the window.

19. (A) 3

Note that every time the first gear moves by one tooth, one tooth is moved on the second gear, which in turn moves one tooth on the third gear, which finally moves one tooth on the last gear. If the first gear moves one revolution, then it moves by 30 teeth, so the last gear moves 30 teeth as well. The last gear has 10 teeth, so it makes 3 revolutions when it moves by 30 teeth.

20. (C) 

Below a regular octagon with three axes representing the folding lines is shown.

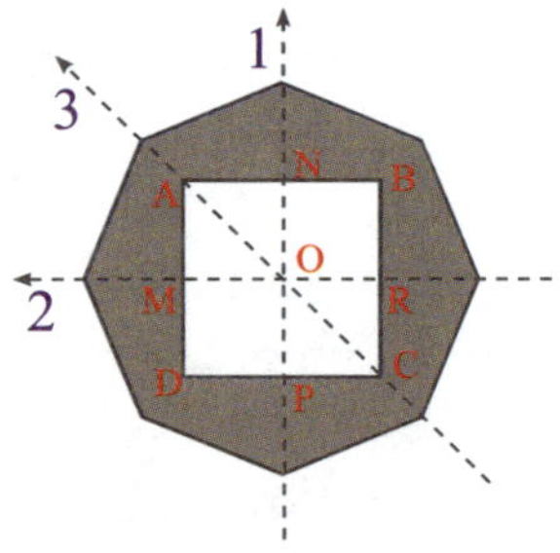

The vertical axis is the 1st folding line, the horizontal axis is the 2nd folding line, and the diagonal axis is the 3rd folding line. NA is the cutting line and we want to see what happens to it when the paper is unfolded. The unfolding process is the reverse of the folding process, so we start with the 3rd folding line. MA is the symmetric image of NA with respect to the 3rd folding line. The segments PD and MD are the symmetric images of NA and MA with respect to the 2nd folding line. The segments NB, RB, RC, and PC are symmetric images of NA, MA, MD, and PD with respect to the 1st folding line. Thus, the unfolded paper is cut along the perimeter of the square ABCD and it looks like the picture (C).

21. (B) There is more wine than vinegar and water together.
Consider the amount of wine as a unit. Then, the amount of wine is 1, the amount of vinegar is $\frac{1}{2}$, and the amount of water is $\frac{1}{3}$. The arithmetical meaning of the answer choices is as follows:
(A) $\frac{1}{2} > 1$ (B) $1 > \left(\frac{1}{2} + \frac{1}{3}\right)$ (C) $\frac{1}{2} > \left(1 + \frac{1}{3}\right)$ (D) $\frac{1}{3} > \left(\frac{1}{2} + 1\right)$ (E) $\frac{1}{2} < \frac{1}{3}$ and $\frac{1}{2} < 1$
(A), (C), and (D) are false statements. (B) is true since $\frac{1}{2} + \frac{1}{3} = \frac{5}{6}$ is less than 1.
In statement (E), "There is less vinegar than either water or wine," both parts need to be true: "There is less vinegar than water and there is less vinegar than wine." This is not true.

22. (D) D

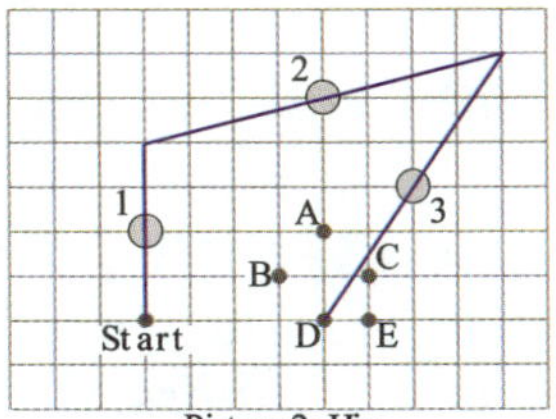

Picture 2: Hip

Construct segments on the grid, paying attention to the vertical and horizontal distances from midpoint to the end points. The last segment will end at point D.

23. (D) 7.5
8 was the most common age and 4 children were 6 years old, so at least 5 children were 8 years old. The smallest number of children of age other than 8 was 4 (6 years old) + 1 (7 years old) + 1 (9 years old) + 1 (10 years old) = 7. 7 + 5 = 12, and that is how many children were at the birthday party, so there were exactly 5 children who were 8 years old. Also, there was exactly one child of each age other than 6 or 8.
The average age of the group was

$$\frac{4 \times 6 + 1 \times 7 + 5 \times 8 + 1 \times 9 + 1 \times 10}{12} = \frac{90}{12} = \frac{15}{2} = 7.5$$

24. (B) 30

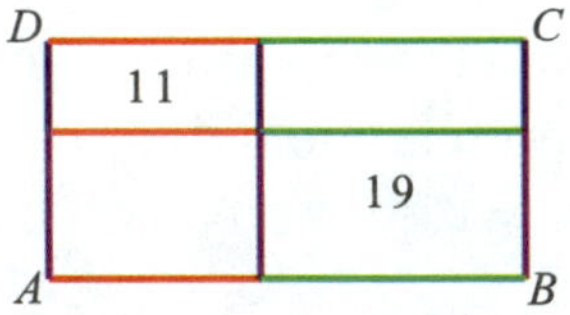

The perimeter of rectangle *ABCD* is made up of segments that correspond exactly to the lengths and widths of two rectangles at opposite vertices. The smallest and largest of these rectangles are at opposite vertices, and we know that their perimeters are 11 and 19. Thus, the perimeter of rectangle *ABCD* is 11 + 19 = 30.

25. (D) 8 and 10

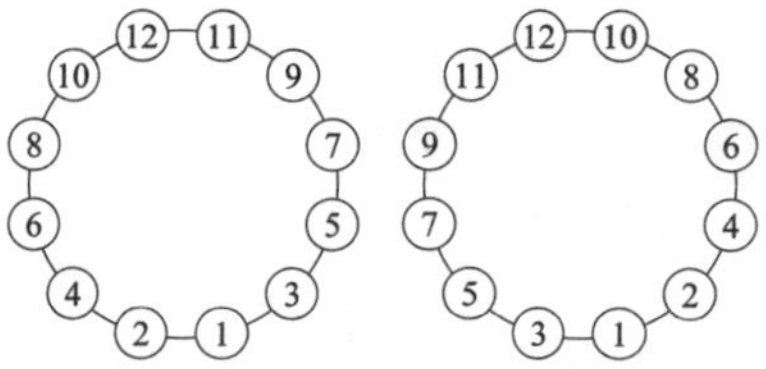

The neighbors of 12 can only be 11 and 10, as there is no 13 or 14. We can place 12 as the first number anywhere, so let's place 12 towards the top. Next to 12 on one side is 11 and on the other side is 10. Moving down from 12 in both directions:
(i) below 11 you can only have 9 since 10 and 12 have already been placed, and then below 10 you can only have 8 since 9, 11, and 12 have already been placed;
(ii) below 9 you can only have 7 since 8, 10, and 11 have already been placed, and then below 8 you can only have 6 since 7, 9, and 10 have already been placed;
(iii) below 7 you can only have 5 since 6, 8, and 9 have already been placed, and then below 6 you can only have 4 since 5, 7, and 8 have already been placed.
Finish this reasoning fill in the circle. The only other way to start is to place 11 to the left of 12 and 10 to the right. In either case, notice that opposite numbers add up to 13. The completed circles are shown above. From the options of given pairs only 8 and 10 are neighbors.

26. (B) 5

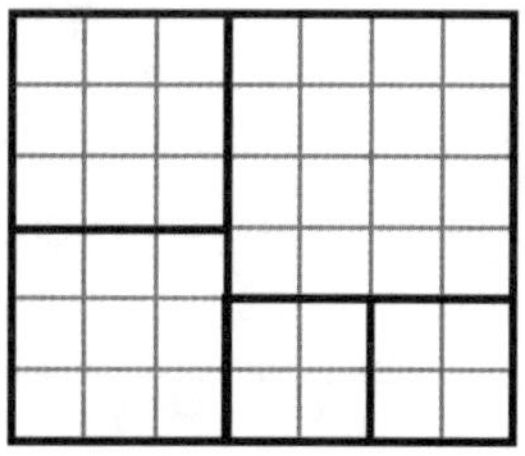

A division into 5 squares is shown to the left. We have to make sure that a smaller number of squares can't cover the 6 × 7 rectangle. The biggest square that fits into the rectangle is a 6 × 6 square. If we use it, then it covers two corners of the rectangle and keeps a 6 × 1 column uncovered. Covering the whole figure requires 6 more unit squares, so 7 squares are required if one square covers two corners of the rectangle.

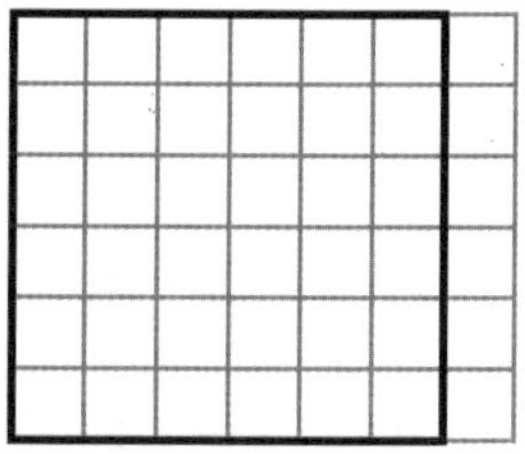

If we use only squares of size 5 × 5 or smaller, then each corner of the rectangle has its own square, so there are already 4 squares covering the 4 corners. Let's look at any rectangle with four squares at its four corners. a × a is the size of the square at the vertex *A*. Similar notation is used for the other vertices (look at the picture to the left). w is the width and h is the height of this rectangle. If the four squares cover the rectangle without any overlap, then $a + d = h = b + c$, so $a + b + c + d = 2h$.

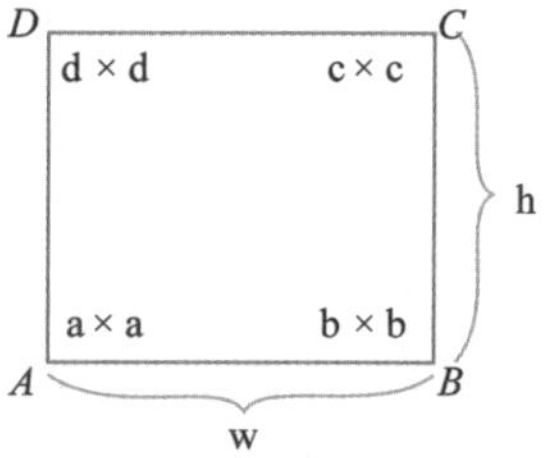

Similarly, $a + b = w = c + d$, so $a + b + c + d = 2w$.
Hence, $2h = 2w$, which happens only when *ABCD* is a square. The original 6 × 7 rectangle is not a square, so it cannot be covered by 4 squares. Therefore, the smallest number of squares Peter can get is 5.

27. (D)

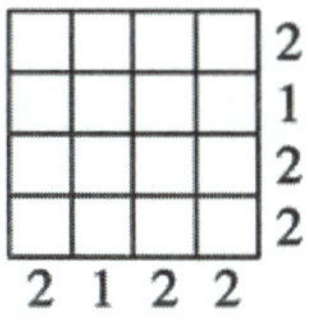

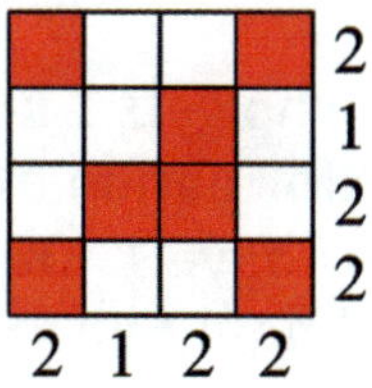

One possible coloring for (D) is shown to the right.
All the other options can be eliminated for different reasons.
(A) indicates that all cells in the top row are red which contradicts the fact that the leftmost column has no red cells, so (A) can't happen.
In (B), count the number of red cells row by row ($1 + 2 + 1 + 3 = 7$) and column by column ($2 + 2 + 3 + 1 = 8$). The numbers are different, so (B) can't happen.
(C) indicates that two bottom rows have no red cells, so the 2nd column can't have 3 red cells.

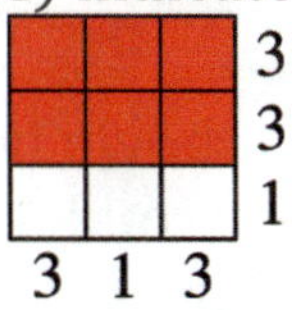

(E) has no red cells in the top row or the first column. By removing the top row and the first column we get a 3×3 table (shown to the left) with the top two rows fully red. Clearly, the middle column has more than 1 red cell.

28. (D) 16 cm^2

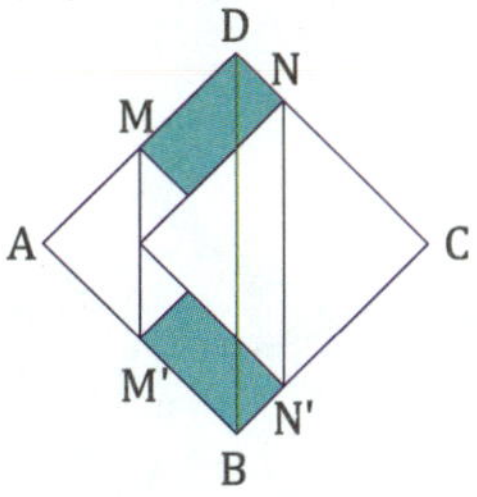

A, B, C, D are vertices of the square and $2d$ is the length of the diagonal BD. The distance from A to BD is d, so the first folding line MM' is $\frac{d}{2}$ units from BD and it is parallel to BD.
Hence, M is the midpoint of DA and $|DM| = \frac{1}{2}|DA|$. Recall that MM' is $\frac{d}{2}$ units from BD, so it is $\frac{d}{2} + d = \frac{3d}{2}$ units from the vertex C.
The second folding line NN' is parallel to the first and exactly in the middle between C and the first folding line, so the distance from NN' to C is $\left(\frac{3d}{2}\right) \div 2 = \frac{3d}{4}$.
Hence, the distance between the diagonal BD and the second folding line NN' is $d - \frac{3d}{4} = \frac{d}{4}$, which is $\frac{1}{4}$ of the distance from C to BD. Hence, $|DN| = \frac{1}{4}|DC|$.
In summary, $|DM| = \frac{1}{2}|DA|$ and $|DN| = \frac{1}{4}|DC|$.
Therefore, the area of the top shaded rectangle is $\frac{1}{2}|DA| \times \frac{1}{4}|DC|$, which is $\frac{1}{8}$ of the area of square ABCD. The bottom shaded rectangle is identical to the top one. Thus, the sum of the areas of the two shaded rectangles is $2 \times \frac{1}{8} \times 64$ cm^2 = 16 cm^2.

29. (C) 5
Let abc be Abid's house number. Then bc is Ben's house number and c is Chiara's house number. A house number never begins with the digit 0, so $1 \le a, b, c \le 9$ and the sum of house numbers, $(100a + 10b + c) + (10b + c) + c$, is 912. Thus, $100a + 20b + 3c = 912$ or $100a + 20b = 912 - 3c$. The last equation is equivalent to $20(5a + b) = 3(304 - c)$, so $304 - c$ is a multiple of 20. It happens only for $c = 4$ since $1 \le c \le 9$. Hence, $20(5a + b) = 900$, which simplifies to $5a + b = 45$ or $b = 45 - 5a$, which simplifies to $b = 5(9 - a)$. The only multiple of 5 between 1 and 9 is 5, so $b = 5$. The equation $5a + b = 45$ becomes $5a + 5 = 45$ and its solution is $a = (45 - 5) \div 5 = 8$. Abid's house number is 854 and 5 is its second digit.

30. (B) 3
Anybody who gets 1 knows that the other person got 2. Ann doesn't have 1 and Bill gains this piece of information from Ann's first statement.
Bill doesn't have 1. Otherwise, he would know Ann's number. Bill doesn't have 2. If he had 2, he would know Ann's number as 3 since 1 is already excluded, so he does not have 2 and Ann now knows it.
So, Ann must have 3 to know Bill's number at this moment. His number is 4 (2 is already excluded), which is a divisor of 20.

Solutions for Year 2014

1. (C) 6
The letter K needs to be rotated twice and the first A needs to be rotated once (as shown in the example). N is sideways, so it needs to be rotated once. G is already correct. The second A is upside-down, so it needs to be rotated twice by 90° to be correct. R is correct, and the O's are written so that they are correct in any position, so they don't need to be rotated. Altogether, Arnold needs to rotate the cards $2 + 1 + 1 + 2 = 6$ times for all of the letters to be correct.

2. (D) 450 g
The weight of the biggest piece is the same as the other 3 pieces, so it is half of the weight of the whole cake. $900 \text{ g} \div 2 = 450 \text{ g}$.

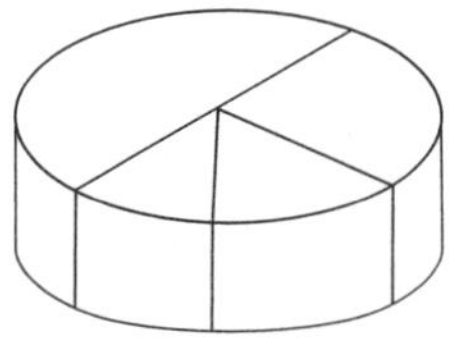

3. (D) 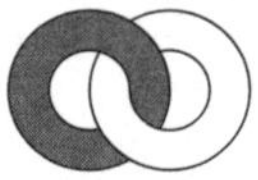

From the front, Peter sees the white ring on the left with the visible overlapping part of that ring in the upper part. From the back, Paul would see that in "reverse"—the white ring on the right, with the upper visible overlapping part of the rings gray, so the answer is (D).

Peter (front)   Paul (back)

4. (A) 0
The ones digits add up to $2 + 3 + 4 = 9$, so nothing is carried to the tens column. The hundreds digits add up to $1 + 1 + 1 = 3$, so nothing needs to be carried from the tens column. Thus, the numbers in their place values already give us a sum of 309, so nothing needs to be added to it. Therefore, the digits replaced by stars are all zeros. $0 + 0 + 0 = 0$.

5. (A) 1
The smallest 5-digit number is 10000 and the largest 4-digit number is 9999. The difference is $10000 - 9999 = 1$.

6. (D) 60 cm
A square with a perimeter of 48 cm has a side length of 48 cm ÷ 4 = 12 cm. This makes the dimensions of the rectangle in the picture 24 cm (twice the side of the square) by 6 cm (half of the side of the square). The perimeter of the rectangle is 2 × (24 cm + 6 cm) = 60 cm.

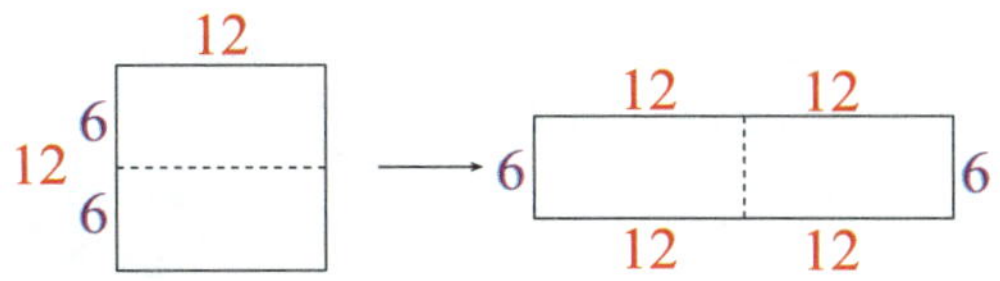

7. (B) 5
Each side of the triangle consists of 6 matches, so Katrina used 6 × 3 = 18 matches to build the triangle. She has 20 matches left to build a square, so each side of the square consists of 20 ÷ 4 = 5 matches.

8. (B) 3

If Alex wants to have 5 dark gray pearls, taking them off from the left end she would have to take 4 white pearls. Taking them off only from the right end, she would have to take 5 white pearls.
She can start by taking one pearl off each end. She can get three more gray pearls by taking one white pearl and one gray pearl off the left end and two white and two gray pearls off the left end. Thus, the smallest number of white pearls she needs to take is 3.

9. (B) the second
Let's find the time it took to complete each lap using the data from the table.

Lap	Duration
1	9:55 to 10:26 = 31 minutes
2	10:26 to 10:54 = 28 minutes
3	10:54 to 11:28 = 34 minutes
4	11:28 to 12:03 = 35 minutes
5	12:03 to 12:32 = 29 minutes

The second lap took the shortest time at 28 minutes.

10. (C) 12:44
There are three possibilities for the first time on Ben's watch: 12:41, 12:43, and 12:47. There are two possibilities for the second time: 12:44 and 12:49. The only change of one minute ahead is 12:43 to 12:44, so it is now 12:44. The options are shown below.

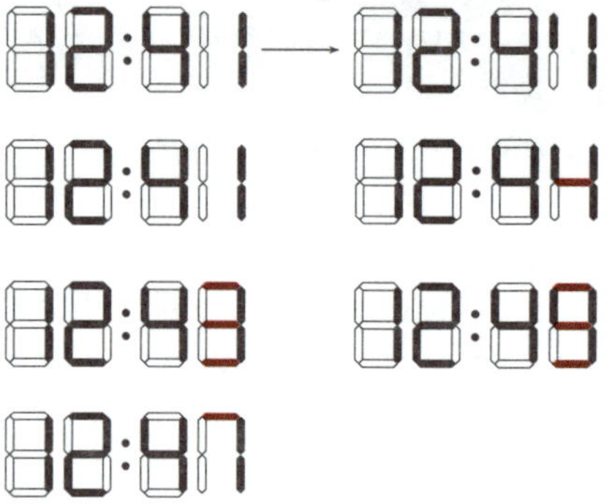

11. (E) It is impossible.
The light gray area consists of $1+\frac{1}{2}+\frac{1}{2}+\frac{1}{2}+\frac{1}{2}+1+\frac{1}{2}+\frac{1}{2}=5$ square units and the dark gray area consists of $\frac{1}{2}+\frac{1}{2}+\frac{1}{2}+\frac{1}{2}+\frac{1}{2}+\frac{1}{2}=3$ square units. Even if we add 1 square unit of dark gray area, the dark gray area will never be as large as the light gray area.

12. (B) 1 km north
Both Henry and John start at the same point (S).
Henry's path (red): 1 km north, 2 km west, 4 km south, and 1 km west ends at point H.
John's path (green): 1 km east, 4 km south, and 4 km west ends at point J. John still needs to walk 1 km north to reach the point where Henry ended up.

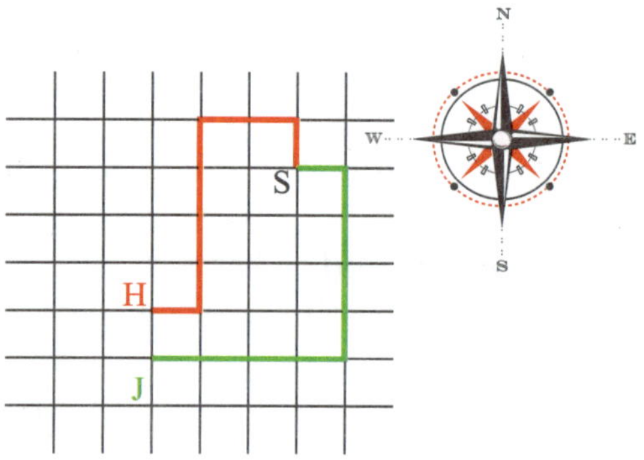

13. (D) 10
7 children eat ice cream every day and 9 children eat ice cream every other day. Yesterday, 13 children ate ice cream, which is the 7 children who eat it every day and 6 of the children who eat it every other day. Today, the 7 children who eat ice cream every day are joined by $9-6=3$ other children who eat it every other day, so $7+3=10$ children will eat ice cream today.

14. (B) B

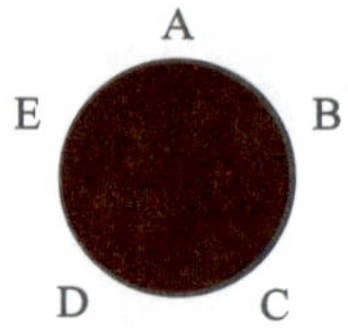

Look what happens with the clockwise order, starting always with A, when one particular kangaroo does not move.
If A doesn't move, then B switches places with C and D with E. The order is ACBED.
If B doesn't move, then A switches places with E and C with D. The order is AEBDC.
If C doesn't move, then A switches places with B and D with E. The order is ACEDB.
If D doesn't move, then A switches places with E and B with C. The order is AECBD.
If E doesn't move, then A switches places with B and C with D. The order is ADCEB.
We are looking for the clockwise order AEBDC. This happens if kangaroo B does not move.

15. (B) B
Each of the pieces can be obtained from a fixed square of area U, by adding/removing a fixed circular segment (or segments) each with an area of M.

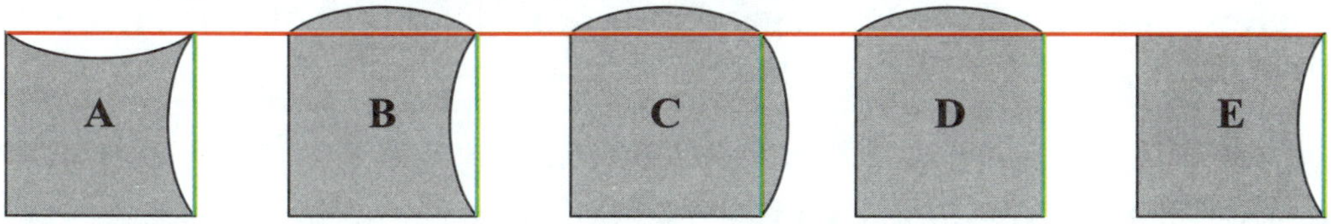

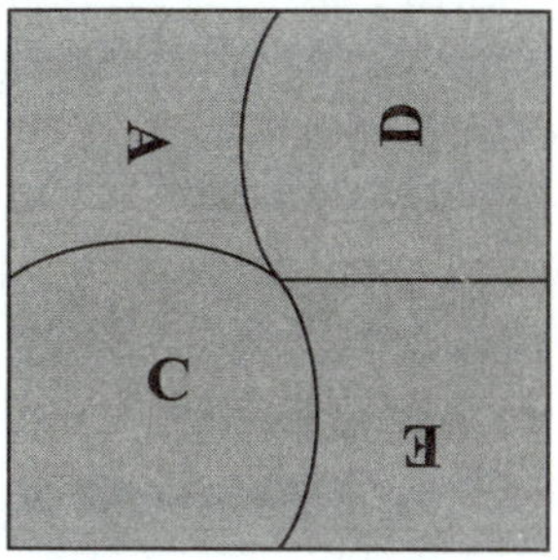

The areas of the pieces (in order shown above) are: U – 2M, U, U + 2M, U + M, and U – M.
The combined areas of four pieces are (in order: without A, without B, without C, without D, without E): 4U + 2M, 4U, 4U – 2M, 4U – M, and 4U + M.
The area of the square we are trying to form is 4U, so we have to exclude piece B and assemble the other four pieces. We can rotate three pieces and match them to the fixed piece C and to each other as shown on the right.

16. (D) 17
The prime factorization of $135 = 3 \times 3 \times 3 \times 5$, so there is only one way to write 135 as the product of three digits. $135 = 3 \times 5 \times 9$, and the sum of these digits is $3 + 5 + 9 = 17$.

17. (A) 4
The tables that have 3 or 4 chairs can accommodate 36 people. The rest of 72 people must be accommodated by tables with 6 chairs. $72 - 36 = 36$ and $36 \div 6 = 6$, so there are 6 tables that have 6 chairs. The other $16 - 6 = 10$ tables have 3 or 4 chairs and can accommodate 36 people. 10 tables with 3 chairs can accommodate only 30 people. If we replace 1 table that has 3 chairs with a table that has 4 chairs, then we can accommodate 1 more person, so to accommodate $36 - 30 = 6$ people we have to replace 6 tables that have 3 chairs with the same number of tables that have 4 chairs. We are left with $10 - 6 = 4$ tables that have 3 chairs. There are 4 tables with 3 chairs, 6 tables with 4 chairs, and 6 tables with 6 chairs.

18. (D) 16

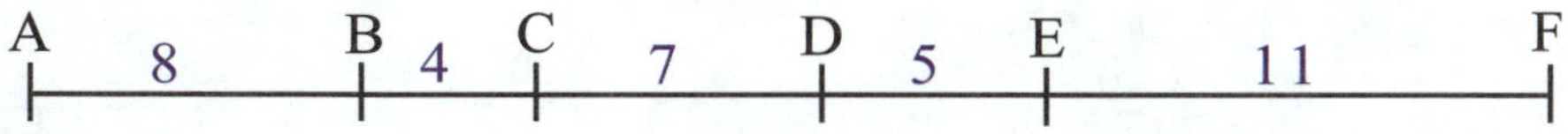

We calculate the distances in the order from F to A.
AC = 12, CE = 12, and AF = 35, so EF = AF – AC – CE = 35 – 12 – 12 = 11.
DF = 16 and EF = 11, so DE = DF – EF = 16 – 11 = 5.
CE = 12 and DE = 5, so CD = CE – DE = 12 – 5 = 7.
BD = 11 and CD = 7, so BC = BD – CD = 11 – 7 = 4.
AC = 12 and BC = 4, so AB = AC – BC = 12 – 4 = 8.
Finally, the distance BE = BC + CD + DE = 4 + 7 + 5 = 16.

19. (E) 13

The number of Marisa's stones without 2 of them is a multiple of 3 and a multiple of 5, so it's a multiple of 15. Hence, the total number of Marisa's stones is 2 + a multiple of 15. In order to not have any stones left when arranging them in groups of 3 and in groups of 5, she needs to reach the next multiple of 15. $15 - 2 = 13$, so she needs 13 more stones.

20. (A) 1

Each pair of faces numbered 1 and 6, 1 and 5, and also 5 and 6 has a common edge, so the faces numbered 1, 5, and 6 have a common vertex. Each pair of faces numbered 1 and 2, 2 and 6, and again 1 and 6 has a common edge, so the faces numbered 1, 2, and 6 have a common vertex. In addition, faces numbered 6 and 4 have a common edge. The number on the face opposite the face with number 4 is 1.

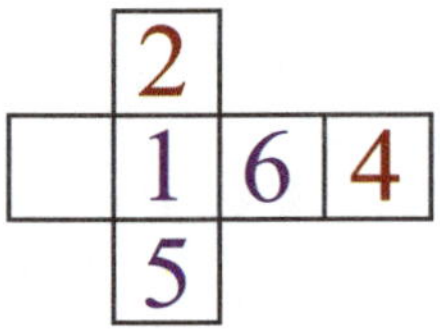

21. (D) 7

To see the image when looking at the cube from the right and the front, the top small cubes of the middle "crossed" rows need to be removed. To see the image from the top, the middle column in the front needs to be removed. Altogether 7 small cubes need to be removed.

22. (A) A

3 min	2 ½ min	2 min	1 ½ min	4 min
A	B	C	D	E

The 5 songs are playing in a loop without any breaks, so it takes $3 + 2½ + 2 + 1½ + 4 = 13$ minutes to play through all of them. Before one hour is completed, the same spot is reached after 13, 26, 39, and 52 minutes. After exactly one hour, the song playing is 8 minutes after the starting point. We would like to know which song is playing 8 minutes after song C. If song C was just starting when Andy left home, then 7½ minutes later song A would start playing ($7½ = 2 + 1½ + 4$). If song C was at the end when he left home, we would be 2 minutes further into the loop, but it would still be song A playing.

23. (E) 29

For number 5 the sum of numbers in adjacent cells has to equal 9. If he placed 5 in the center cell, then the sum of the numbers adjacent to it would be the sum of the four numbers not yet placed, 6, 7, 8, and 9. Clearly, that is more than 9. If the number 5 was between 1 and 3, and the next smallest available number 6 was placed in the center, the sum would already also exceed 9 ($1 + 3 + 6 = 10$). Therefore, the only option of placing 5 would be between 1 and 2, with the number 6 in the center. The sum of the numbers in the cells adjacent to the cell with the number 5 is now $1 + 2 + 6 = 9$. With number 6 in the center, the numbers in the cells adjacent to it are 5, 7, 8, and 9 and their sum is $5 + 7 + 8 + 9 = 29$ (the exact placement of 7, 8, and 9 does not matter).

1		3
5	6	
2		4

24. (C) 20

There are 60 trees in total. Since every other tree is a maple, there are 30 maple trees. Now, every third tree is either a linden or a maple, and already every other tree is a maple, so every sixth tree has to be a linden. Hence, there will be 10 linden trees. The remaining trees are birches, so there are $60 - 10 - 30 = 20$ of them.

25. (E) 

The cube is transparent, so the ribbon on the side facing you and the side opposite will be visible from a given perspective.

(A)  (B)  (C)  (D)  (E) 

The picture in (A) shows the transparent cube from the front (or back). The picture in (B) shows it from the right (or left) side. The pictures in (C) and in (D) show the cube from the top (or bottom) depending how it would be rotated (by 90°). Thus, the four pictures from (A) to (D) show the cube from all six perspectives.
The picture in (E) does not show the cube from any of the six sides. There would have to be a piece of ribbon somewhere on the front or back face in order for this to be a view of the cube.

26. (D) 90 min

One hour after leaving the castle the king is 5 km away from the castle, and the first messenger is sent back. It will take this messenger half an hour to get back to the castle (90 minutes from the start of the journey). Two hours after leaving the castle the king is 10 km away, and the second messenger is sent back. It will take this messenger one hour to get back (3 hours from the start of the journey). Three hours after leaving the castle the king is 15 km away, and the third messenger is sent back. It will take this messenger an hour and a half to get back (4 ½ hours from the start of the journey). The time interval between any two consecutive messengers arriving at the castle is an hour and a half, which is 90 minutes.

27. (B) either 7 or 8

The product of 3 and the two digits not erased is 36, so the product of the two digits not erased is $36 \div 3 = 12$. The two digits not erased are either 2 and 6 or 3 and 4.
In the first case the erased digit was $15 - (2 + 6) = 7$ and in the second case the erased digit was $15 - (3 + 4) = 8$. Thus, the options for the erased digit are either 7 or 8.

28. (C) 2

The number of days Peter Rabbit ate 9 carrots only cannot exceed 3 since $4 \times 9 = 36$ is more than 30. It cannot be 3 days since $30 - 3 \times 9 = 3$ carrots would not be enough for the days when he eats 1 cabbage and 4 carrots only. 2 days does work since $30 - 2 \times 9 = 12$ and 12 carrots can be eaten on 3 days when Peter Rabbit eats 1 cabbage and 4 carrots only.
1 day doesn't work since $30 - 1 \times 9 = 21$ is not a multiple of 4. Similarly, $30 - 0 \times 9 = 30$ is also not a multiple of 4. Hence, there were 2 days when Peter Rabbit ate only 9 carrots per day and 3 days when he ate 1 cabbage and 4 carrots only.
During these $2 + 3 = 5$ days Peter Rabbit ate 3 cabbages, which is 6 cabbages short for his total of 9 cabbages, so on 3 days he ate 2 cabbages only.
During these $5 + 3 = 8$ days Peter Rabbit ate: $2 \times 9 + 3 \times 4 = 30$ carrots and $3 \times 1 + 3 \times 2 = 9$ cabbages. On the other $10 - 8 = 2$ days he ate only grass.

29. (C) 4 days

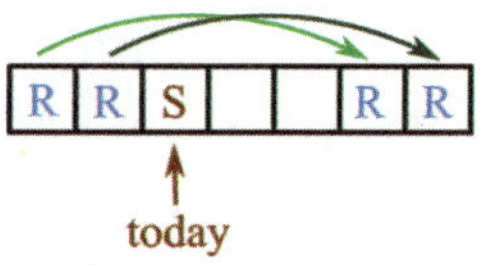

The table shows the rules for the weather in Fabuland: each sunny day is immediately preceded by two consecutive rainy days, and five days after a rainy day it is another rainy day. What happens for two days after a sunny day is that it rains as well, since if either of these days were sunny it would have had to be preceded by two consecutive rainy days, which cannot happen. So, after a sunny day, there will be two rainy days.
We cannot predict what will happen five days from today. It could be sunny, with two consecutive rainy days immediately preceding and following it, and the pattern could continue. Or, five days from now it could rain, which would actually mean rain forever.

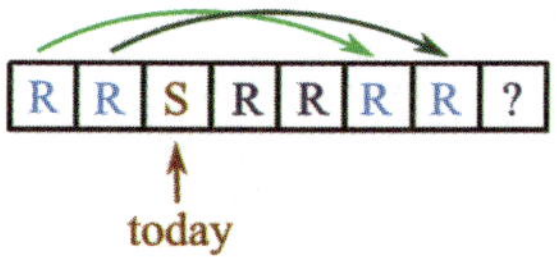

Therefore, we can only predict the weather with certainty for 4 days after today.

30. (E) 23

Alice is the oldest and there are 10 grandchildren, all of different ages. For Alice's age to be the youngest possible, the other ages need to be the greatest possible, that is, as close to Alice's age as possible. To try different numbers, we should start with an age that is slightly more than $180 \div 10$.
If Alice is 20 years old, then the largest possible sum of all the grandchildren's ages is
$20 + 19 + 18 + 17 + 16 + 15 + 14 + 13 + 12 + 11 = 155$.
If Alice is 21 years old, then the largest possible sum is
$21 + 20 + 19 + 18 + 17 + 16 + 15 + 14 + 13 + 12 = 165$.
If Alice is 22 years old, then the largest possible sum is
$22 + 21 + 20 + 19 + 18 + 17 + 16 + 15 + 14 + 13 = 175$.
If Alice is 23 years old, there are many possibilities for all the ages to add up to 180. For example, $23 + 22 + 21 + 20 + 19 + 18 + 17 + 16 + 15 + 9 = 180$. Since 23 is the smallest age for which we can get such a sum, 23 is the youngest Alice can be.

Solutions for Year 2016

1. (C) 

The sign in (C) has two lines of symmetry, one horizontal and one vertical.

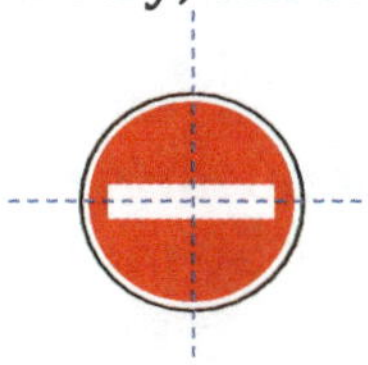

Other signs with lines of symmetry are (A), with only a horizontal one, and (E), with a vertical one.

2. (E) a twelfth
Since each quarter was cut into 3 pieces, there are now $3 \times 4 = 12$ pieces. Thus, each piece is a twelfth of the whole pizza.

3. (A) 2 cm, 3 cm, 5 cm
The thread is folded into ten pieces, so they are each 1 cm long. The first part (marked with red colors) consists of three pieces, so it is 3 cm long. The middle part (marked with grays) is 5 cm long, and the last part (marked with blues) is 2 cm long.

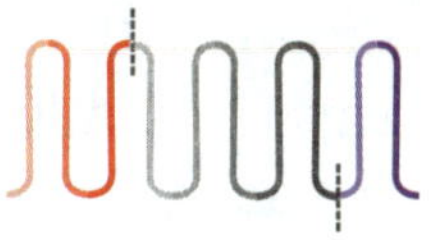

4. (C) 4
Both cards with flowers are held by only one magnet, so these two magnets must stay in place. The picture shows two groups of overlapping cards. The first group (on top) has only two cards. One of the magnets is holding both of them in place, so the other magnet can be removed (it is colored red here). The second (bottom) group consists of three mutually overlapping cards kept in place by one magnet, as well as two other cards. Because these two other cards do not overlap, at least two more magnets are needed to keep them in place, one for each card. There are six different choices for which magnets can be removed from the bottom group. One option is shown in red to the right. Thus, a total of 4 magnets are necessary to keep all seven cards in place, one for the top group and three for the bottom group.

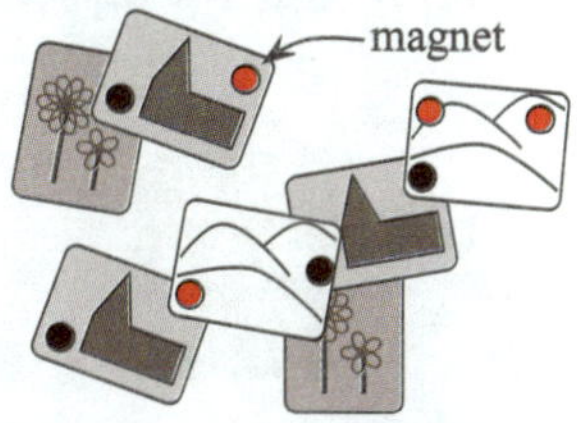

5. (E) 50 cm^2
The larger square has an area of $10 \text{ cm} \times 10 \text{ cm} = 100 \text{ cm}^2$. The green square is made up of 4 triangles, each of which is half of one of the four squares which make up the larger square. So, the smaller green square has an area of half the larger square, or 50 cm^2.

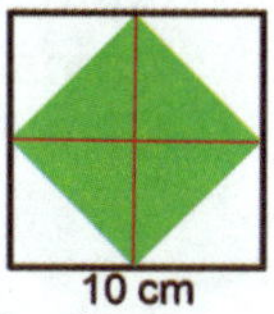

6. (B) 2

After interchanging the two utensils between the two plates on the left (shown in red) and two utensils next to the plate on the right (shown in blue), the arrangement becomes pleasing to Alice's mother.

7. (D) 50

The 25 pairs of shoes will go on 50 of the centipede's feet. Because it still has $100 - 50 = 50$ feet without shoes, the centipede needs to buy 50 more shoes.

8. (C) 4

Tom's box is six cubes long, two cubes high, and two cubes deep, so it is made out of $6 \times 2 \times 2 = 24$ cubes. John's box has 6 cubes on one level, so to make a box with 24 cubes he will need $24 \div 6 = 4$ levels. Hence, John's box will have four levels.

9. (C) 2

There are two pairs, Bea & Pia and Mary & Karen, identified by their initials in the figure below. Bea & Pia are facing each other, and Mary & Karen have their backs to each other. They are drawn with their feet showing which direction each is facing.

In each pair, one girl is sleeping on her right side and one on her left. Hence, 2 girls are sleeping on their right sides.

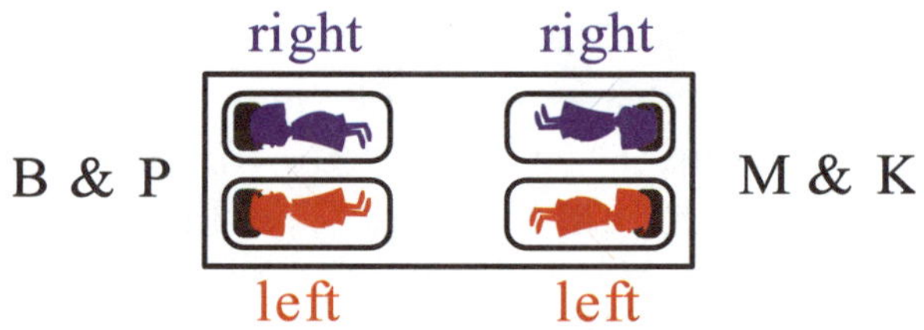

10. (B) B

Start by folding face A up. Then, fold along the other dotted line at face B. This will bring faces C, D, and E up. Then fold along all the remaining dotted lines. All the faces will be adjacent to B, so B will be on the bottom, that is, opposite the open side.

11. (A)

For the figure in (A), the top triangle is half of the square whose diagonal (shown in red) is the same as the side of the bottom square, so the square with the red diagonal is smaller than the bottom square. Another square of the size of the bottom square fitting to the right angle of the top triangle extends beyond the figure in (A) as shown by the blue lines. Hence, it is impossible to make the figure in (A) by gluing the two identical squares.

All of the other figures can be formed by gluing the two identical squares.

(A)

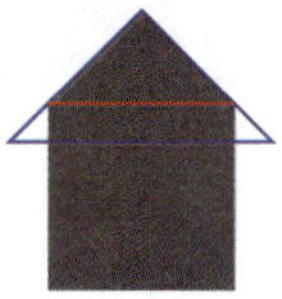

(B)

(C)

(D)

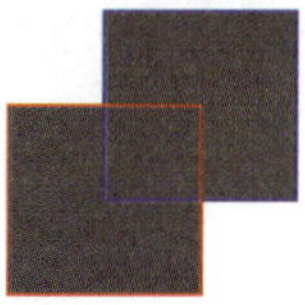

(E)

12. (C) 3

The one day when Ann has a day off, Mary and Nata come to work. The other four days Ann works with one of the women, two days with Mary and the remaining two days with Nata. Thus, Nata works 3 days per week.

13. (C) *C*

Each step below shows a movement of all squirrels.

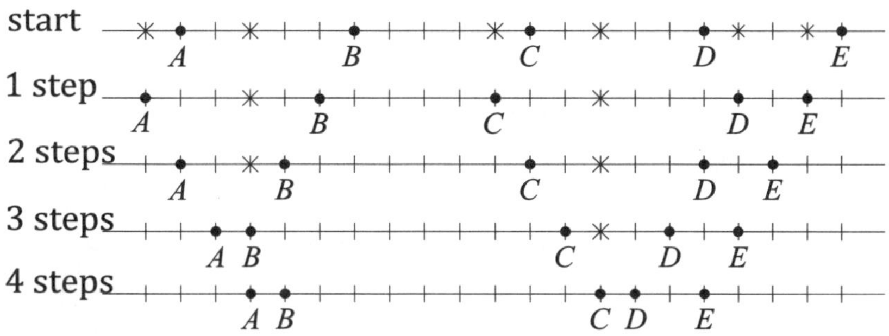

In the first step, squirrels *A*, *C*, *D*, and *E* each pick up one nut. None of them picked up a nut in the second step, but *B* gets a nut in the third step. At this point, each squirrel has picked up one nut, and there is only one nut left. Squirrel *C* has the shortest distance to it, so squirrel *C* gets two nuts.

14. (D) 10

Since we know that only half of the girls are sitting with a boy, there must be twice as many girls as boys, so in a class of 30 students there are 10 boys.

15. (B) 2975

If we cut the strip, 2581953764, between the 8 and the 1 and also between the 3 and the 7, then the sum is $258 + 1953 + 764 = 2975$. If one piece has 5 or more digits, then the sum is greater than 10,000. Hence, we consider only cuts such that each piece has at most 4 digits. There is only occurrence of 1, so if there are two 4-digit pieces, then the sum is greater than $1{,}000 + 2{,}000$, which is more than 2975. There is exactly one 4-digit piece because there are 10 digits on the strip of paper. If the 4-digit number starts with 3 or more, the sum is greater than 3,000. If it starts with 2, then the only option is $2581 + 953 + 764$, which exceeds 3,000. Therefore, the 4-digit piece must start with 1, producing the sum $258 + 1953 + 764 = 2975$.

16. (E) 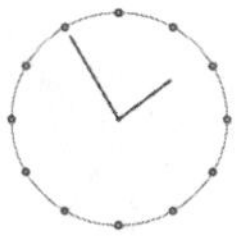

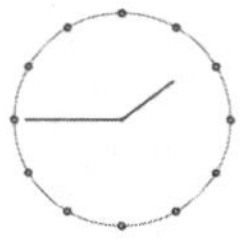

The mirror image of the clock looks like it is showing 1:45, so the clock must actually be showing 10:15 (shown to the right). Ten minutes earlier the clock would have been pointing to 10:05, which would have looked like 1:55 in the mirror.

17. (A) 8

If she can feed 4 cats for 12 days, then grandmother has $4 \times 12 = 48$ daily portions. Divided among six cats, these 48 portions will last $48 \div 6 = 8$ days.

18. (D) 5

In the number represented by BENJAMIN each digit appears once, except for the digit corresponding to N, which is repeated. The sum of the digits without one of the N's is $1 + 2 + 3 + 4 + 5 + 6 + 7 = 28$.

For a number to be divisible by 3, the sum of its digits must also be divisible by 3, so the sum $B + E + N + J + A + M + I + N = 28 + N$ must be divisible by 3. For the available digits, only $28 + 2$ and $28 + 5$ are divisible by 3. The digit 2 is even, so it can't be used for N because N as the last letter in the word BENJAMIN must be odd, leaving 5 as the only option for N.

19. (A) 53

If we add 3 to Carl's age, and then add the result to the ages of the triplets, then the total would have to be a multiple of 4. Testing the options gives us:

(A) $53 + 3 = 56 = 4 \times 14$.

(B) $54 + 3 = 57$ is not a multiple of 4.

(C) $56 + 3 = 59$ is not a multiple of 4.

(D) $59 + 3 = 62$ is not a multiple of 4.

(E) $60 + 3 = 63$ is not a multiple of 4.

Thus, (A) is the only valid option.

Each of the triplets is 14 years old and Carl is 11 years old. The sum of their ages is 53.

20. (C) 40 cm

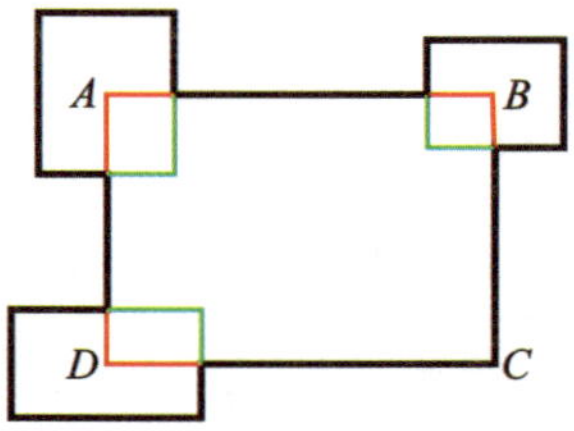

When computing the length of the thick black line, the red segments are missing from the perimeter of *ABCD* and the green segments are missing from the perimeters of the rectangles centered at the vertices *A*, *B*, and *D*, so the length of the thick line is the difference between 30 cm + 20 cm and the combined length of all the red and green segments. At *A* the two red and two green segments form a rectangle with each side being half of the corresponding side of the bigger rectangle centered at *A*, so the combined length of these two red and two green segments is half of the perimeter of the rectangle centered at *A*. The same is true for vertices *B* and *D*, so the combined length of all the red and green segments is $\frac{1}{2} \times 20$ cm $= 10$ cm. Therefore, the length of the thick line is 30 cm + 20 cm – 10 cm = 40 cm.

21. (D) 

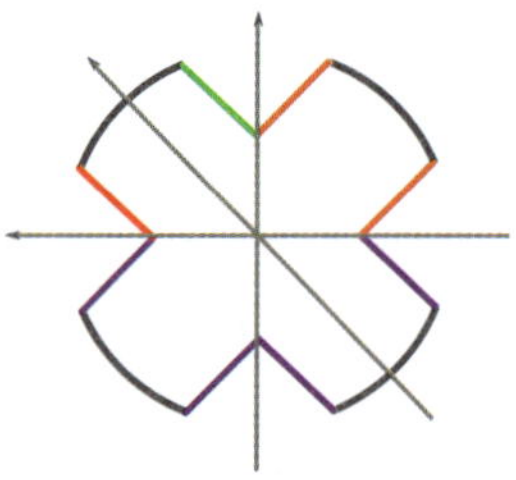

The horizontal axis is the 1st folding line, the vertical axis is the 2nd folding line and the diagonal axis is the 3rd folding line. The cutting line is parallel to the 3rd folding line. It is shown to the right as a red segment and we want to see what happens to it when the paper is unfolded.

The unfolding process is the reverse of the folding process, so we start with the 3rd folding line.

The green segment is a symmetric image of the red segment with respect to the 3rd folding line. The two orange segments are symmetric images of the red and green segments with respect to the 2nd folding line. The four blue segments are symmetric images of the previous four segments with respect to the 1st folding line. These 8 segments are the cutting lines of the unfolded paper and this unfolded paper looks like the figure in (D).

22. (B) 5
To write almost all of the numbers possible, Richard can start with a number made of only the digit 1 repeated 5 times, and then shorten it by decreasing the number of digits and increasing the value of the last digit. This gives him the numbers 11111, 1112, 113, and 14. Since $4 = 2 + 2$, the last valid number is 122. Altogether, Richard writes 5 numbers.

23. (D) 6
Six such shapes can be cut out as shown to the right using six different colors. Seven shapes (or more) are impossible since each shape requires four small squares, and seven shapes made of four squares each would need $7 \times 4 = 28$ small squares, and only $5 \times 5 = 25$ small squares are available.

24. (B) 10
6 more chairs are needed to have 4 chairs at every single table. We can arrange all chairs so that one single table has 2 chairs, another has no chairs, and all the others have 4 chairs each. We want all the single tables, including two tables with missing chairs, to convert into double tables with 6 chairs each. Begin with four regular single tables. From each of them take away 1 chair to form two double tables with 6 chairs each. Add the 4 chairs taken away to the two tables with missing chairs and convert these tables into one double table with $2 + 4 = 6$ chairs. Now we have three double tables (from six singular tables) with 6 chairs each but we need more tables to get a surplus of 4 chairs in the end. That involves four more regular single tables and converting them into two double tables (each with 6 chairs) which leaves 4 chairs for the surplus. Altogether, there are five double tables (each with 6 chairs) and the surplus of 4 chairs. Thus, Luigi has 5 double tables, so he got $5 \times 2 = 10$ single tables from Giacomo.

25. (B) 9
To make a triangle built from identical small triangular tiles, start with one tile at a vertex. Then, adding one row at a time, we see that the next larger triangle would require 3 additional tiles, the one after that 5 more tiles, and next one after that 7 more, then 9, 11, 13, etc. So, the total number of tiles required is 1, $1 + 3 = 4$, $1 + 3 + 5 = 9$, $1 + 3 + 5 + 7 = 16$, etc.
Clara has already put seven tiles together, so the first option would be to put two additional tiles to have a triangle with 9 tiles. This cannot be done for Clara's figure. Putting 9 additional tiles, however, will allow for construction of the large triangle with $7 + 9 = 16$ tiles as shown in the figure to the right.

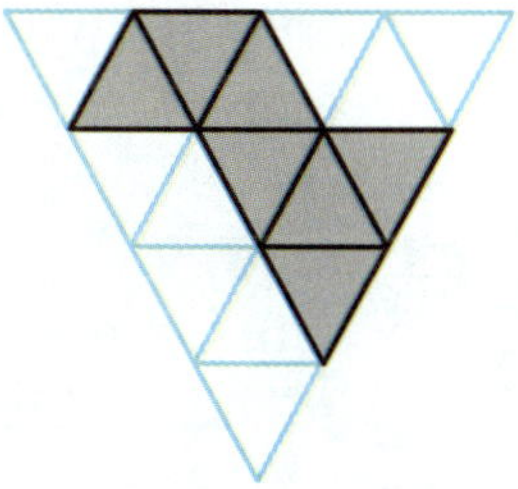

26. (D) 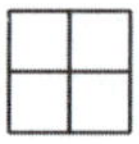

The five given faces of the $2 \times 2 \times 2$ cube are shown to the left. The cube is built from 8 identical unit cubes, and each unit cube is either completely black or completely white. We have to pick the 6th face of the bigger cube from 5 options.

(A)  (B)  (C)  (D) 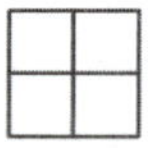 (E) 

Among the five given faces there is no face that has an edge with two adjacent black squares, so the face (C) can't be adjacent to any of the five given faces. The same is true for the face (E). The number of small black squares on all six faces combined must be a multiple of 3. That is because each small square has one vertex that is also a vertex of the big cube. At each vertex of the big cube there is a small cube of one particular color. If one face of the small cube is black, then all faces of this small cube are black. Three of these faces are parts of the three faces of the big cube. Thus, each small black cube contributes 3 small black squares to the surface of the big cube. Among the five given faces, there are 6 black squares, and 6 is a multiple of 3, so the sixth face must have either 0 or 3 black squares. This eliminates options (A) and (B). Since option (E) has already been eliminated, option (D) with all white squares is the only option left. Face (D) must be opposite of the first face among the given faces, which is the only face without a totally white edge. The actual cube is shown to the right.

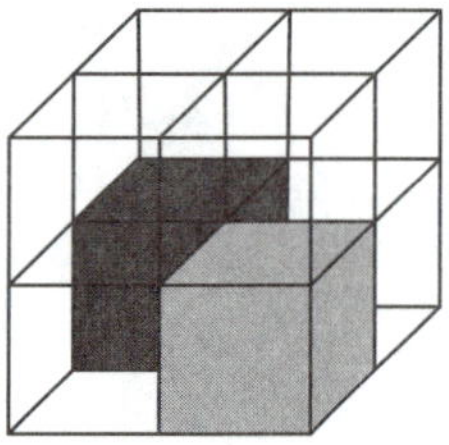

27. (D) 13

The side with the 3 and the side with the 1 share a circle, so the sum of their non-shared circles must be $7 + 3 = 1 + y$, where y is the number in the lower left corner. So, y is $7 + 3 - 1 = 9$. Using the same method for the bottom and lower right sides, we see $9 + 6 = 2 + X$, so $X = 6 + 9 - 2 = 13$. A complete solution is shown to the far right. n is any number, $b = n + 5$ and $a = n + 10$.

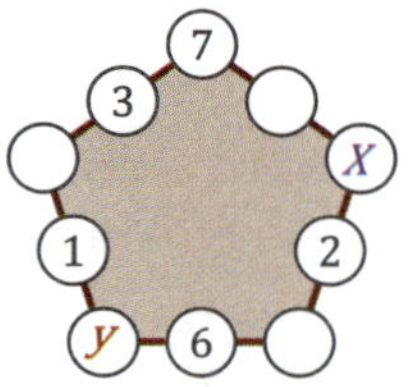

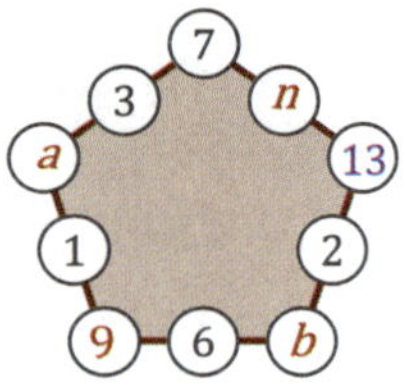

28. (E) 9

Instead of the circle, square, and triangle, consider the letters C, S, and T, respectively. The problem is equivalent to the following five requirements. (i) C, S, T are different digits; (ii) CSC is a 3-digit number; (iii) ST is a 2-digit number; (iv) $C + S + C = 10S + T$ and (v) $S + T = S$. From (v), $T = 0$. Then from (iv), $2C = 9S$. From (i), $C, S > 0$ and C, S are digits. Therefore, the only solution for $2C = 9S$ is $C = 9$ and $S = 2$. Of course, 9, 2, and 0 are different digits. Thus, the circle (letter C) represents the digit 9.

29. (C) 36
Let's look at each of the answers listed.
To get to 12 we can do $12 \times 3 \div 3$. This is performing 2 operations and we still have to do 58 more operations. If we multiply by 3 and divide by 3 29 times each, the number will stay the same and we will have performed all 60 actions. We can apply this to 18, since $12 \times 3 \div 2$ is 18. To get 72 we just need to do $12 \times 3 \times 2$, and to get 108 we have to do $12 \times 3 \times 3$. In each case, we can perform 58 more actions that cancel each other out to stay at this value.
The only value for which we cannot do this is 36; if we take 12 and multiply by 3 we get 36, but that leaves 59 operations, and since this is an odd number we can't repeat it like the others. Therefore, the little kangaroo cannot get 36 with 60 operations.

30. (E) 537
The second 3-digit number begins with an even digit. If this digit is 2, then the last digit of the first 3-digit number is 1. To make each 3-digit number as small as possible we start with 3♣1 and 2♦♥ where 1, 2, 3, ♣, ♦, and ♥ are different digits. The smallest sum of the tens digits is 4 ($4 = 4 + 0$ and $4 = 0 + 4$), so our numbers could be 301 and 245 or 341 and 205, with 546 as the sum.
If the first digit of the second number is 4, then the last digit of the first 3-digit number is 2 and we can use 1 as the first digit of the first 3-digit number. We are adding 1♣2 and 4♦♥, so to make the sum as small as possible use 0 and 3 as the tens digits. The options are $102 + 435$ or $132 + 405$, with 537 as the sum. Using 6 or 8 as the first digit of the second 3-digit number makes the possible sum larger than 700, so 537 is the smallest possible sum.

Solutions for Year 2018

1. (A) 3
The gray lines extend the flight of the arrows.
6 balloons are hit, so $9 - 6 = 3$ are not hit.

2. (C) 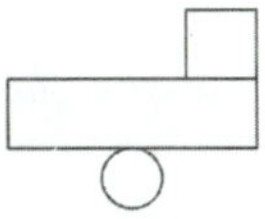

When we look at the table with 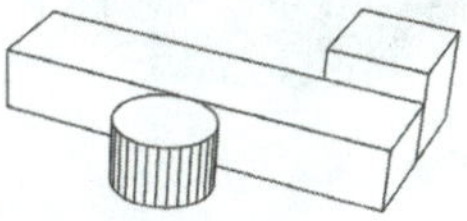 from above, we see a long rectangle. On one side of the rectangle near its center we see a circle and on the other side, to the far right, we see a square.

3. (B) 18
The first time Diana got 14 points by hitting the middle ring twice, so hitting the middle ring stands is worth $14 \div 2 = 7$ points. The second time Diana got 16 points by hitting the center ring once and the middle ring once, so hitting the center ring is worth $16 - 7 = 9$ points. The third time she hit the center twice, so she got $9 \times 2 = 18$ points.

4. (B) B

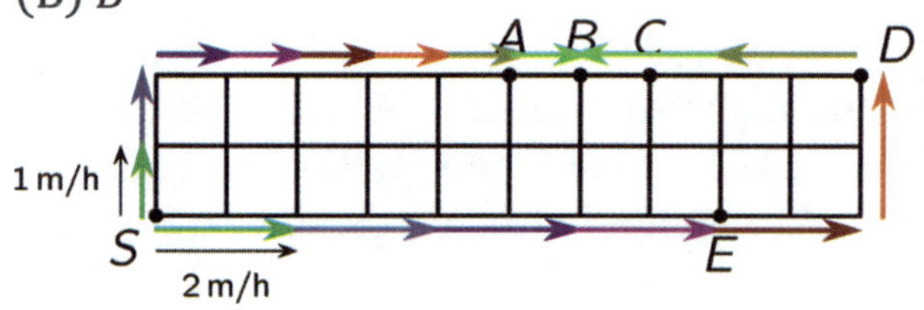

The garden is a 2×10 rectangle, so its perimeter is $2 \times (2 + 10) = 24$ meters long. During one hour the two snails cover 3 meters of the perimeter since $1 + 2 = 3$. They need 8 hours to cover the whole perimeter since $8 \text{ h} \times 3 \text{ m/h} = 24 \text{ m}$. After 2 hours the slow snail reaches the next corner and moves toward the points A, B, C, and D. 6 hours later both snails meet at point B as shown above.

5. (D) 13

The subtraction 

can be replaced by the addition 

The sum is a 2-digit number with 3 as its last digit. The only digit added to 5 that would make the last digit of the sum 3 is 8, so we are adding 25 and 28 with 53 as the result. Indeed, $53 - 28 = 25$. The digits in the painted cells are 5 and 8. Their sum is 13.

6. (E) 72 cm
To get the perimeter of the star from the perimeter of the square, each side of the square can be replaced by two sides of the equilateral triangle, so the perimeter of the star is twice the perimeter of the square or $2 \times 36 \text{ cm} = 72 \text{ cm}$.

7. (D) Saturday
The 3rd day of the month is a Friday, so 3 weeks later it is also a Friday and it is the 24th day of the month since $3 + 3 \times 7 = 24$. The 25th day of the month is the next day, so it is a Saturday.

8. (C) 7
It may happen that the first six consecutive rolls result in six different numbers. Seven rolls must end with at least one repetition of the six possible outcomes, so we have to roll a regular die 7 times to be sure that at least one result will be repeated.

9. (C) 12 cm

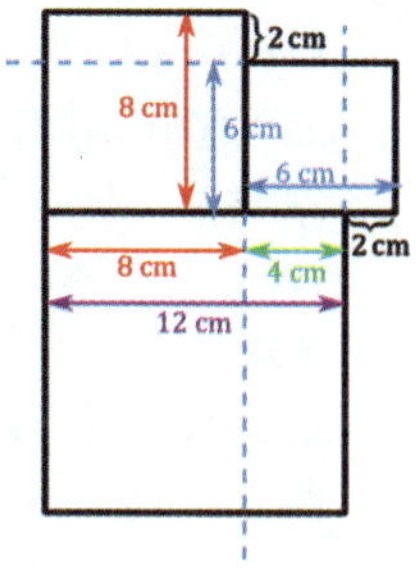

The blue arrows represent the length of 6 cm (the side length of the smallest square). The green arrow represents the length of 6 cm – 2 cm = 4 cm. The red arrows represent the length of 2 cm + 6 cm = 8 cm, so the side length of the biggest square is 4 cm + 8 cm = 6 cm – 2 cm + 2 cm + 6 cm which is 6 cm + 6 cm = 2 × 6 cm = 12 cm or twice the side length of the smallest square.

10. (A) 2

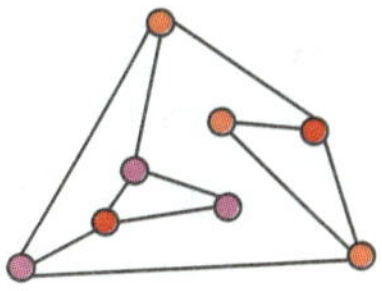

Each bulb has either 2 or 3 neighbors, so touching just one bulb will not light all 8 bulbs. Touching two bulbs, like in the example shown to the right in red, will do that.

11. (E) They are all the same.

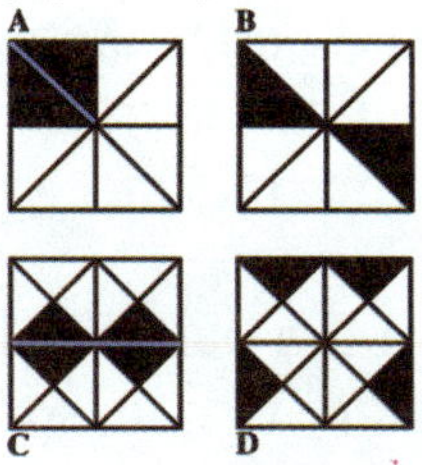

Square A consists of 8 identical triangles. 2 of them are black and 6 of them are white, so the ratio of the black area to the white area is 2 to 6 (or 1 to 3). The same is true for square B.

Square C consists of 16 identical triangles. 4 of them are black and 12 of them are white, so the ratio of the black area to the white area is 4 to 12, which is the same as the ratio 2 to 6. The same is true for square D. Thus, the ratios are the same for all four squares.

12. (B)

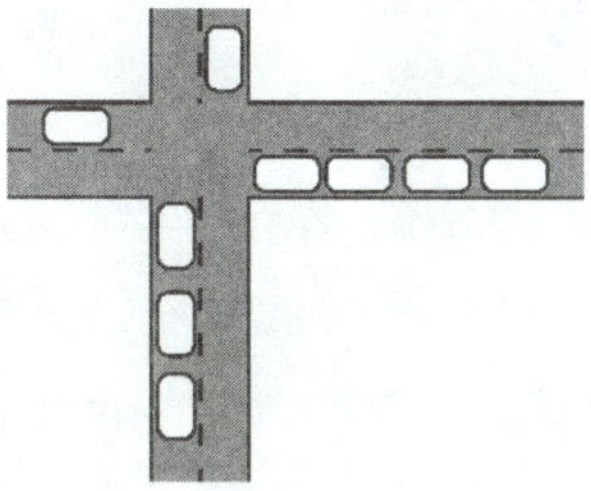

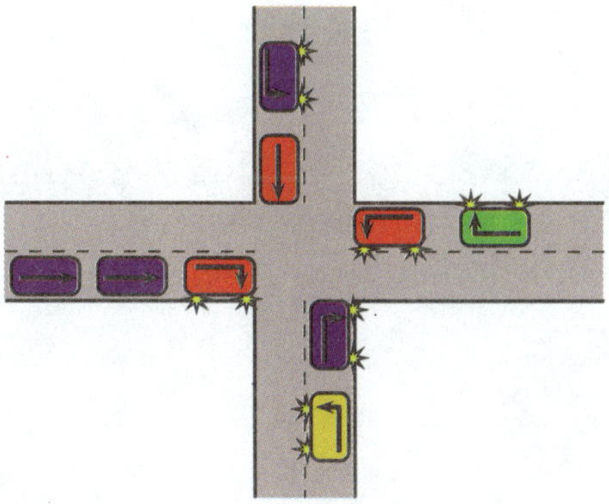

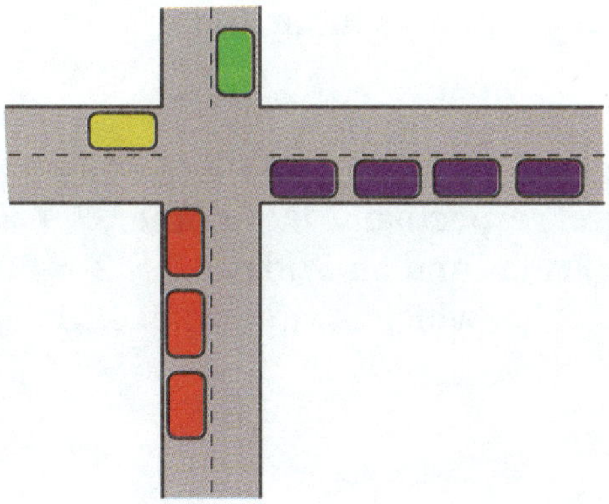

13. (E) 5

The top equation states that "green" + "pink – "blue" = 8, so "green" + "pink" = 8 + "blue," and "green" + "pink" must be at least 9. That can happen only for 4 + 5 or 5 + 4. From the other equation, we see that "green" × "brown" = 8 × "blue." "Green" can't be 5 since 5 is a prime number but it is not a divisor of 8 and it is not a divisor of 1, 2, 3, or 4 either. Thus, the top equation is $4 + 5 - 1 = 8$ and the bottom equation is $4 \times 2 \div 1 = 8$. The number covered by "pink" (the spot with the star) is 5.

14. (A) door 1

"The sum of two and three is five" is a true sentence, so the other two statements are false. Therefore, the lion is not behind door 2, and is behind door 1.

15. (A) Adam

The first throw is from Eva to Adam. Adam throws the ball either to Isaac or to Urban.
If Adam throws the ball to Isaac (the second throw), then Isaac can't return the ball to Adam, so Isaac throws the ball to Urban (the third throw). Urban can't return the ball to Isaac, so he throws the ball to Adam (the fourth throw). Then, the fifth throw is from Adam to Isaac, because Adam can't return the ball or Urban.
If Adam throws the ball to Urban (the second throw), then Urban can't return the ball to Adam, so Urban throws the ball to Isaac (the third throw). Isaac can't return the ball to Urban, so he throws the ball to Adam (the fourth throw). Then, the fifth throw is from Adam to Urban.
In either case Adam will do the fifth throw.

16. (C) 21

The red 2 is necessary since the cell underneath has a common edge with the cell with a 2. The sum 3 + 2 is 5, so the sum of numbers in any two cells with a common edge must be 5. This determines all entries in the table as shown to the right. 2 is repeated 6 times and 3 is repeated 3 times, so the sum of all the numbers in the table is $6 \times 2 + 3 \times 3 = 12 + 9 = 21$.

17. (C) Friday

On Monday, Tuesday, Wednesday, and Thursday 5, 5×2, $5 \times 2 \times 2$, and $5 \times 2 \times 2 \times 2$ people received the picture, respectively. Altogether, $5 + 10 + 20 + 40 = 75$ people received the picture before Friday and on Friday $40 \times 2 = 80$ new people joined them, increasing the total from 75 to 155, which is more than 100 people.

18. (E) 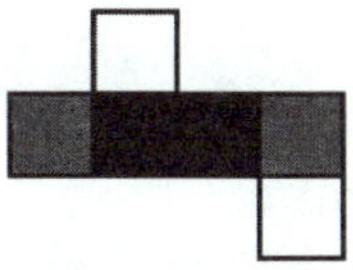

All the five nets shown have 4 squares in one row, so we can fold them in the same way. First, fold along the 3 internal edges of the row of 4 squares, so the 1st & the 3rd and the 2nd & the 4th squares become pairs of opposite faces of the cube. Then, fold the remaining two squares to close the two openings, so these squares become the top and bottom faces of the cube. In (E) these last two faces have the same color, so this net is not appropriate for any cube with opposite faces being of different colors. The other four nets satisfy the requirement.

19. (A) 0
This calculation is equivalent to $(100A + 10B + C) + (100C + 10B + A) = D \times 1111$ or $101 \times A + 101 \times C + 20 \times B = 101 \times 11 \times D$. Each of the four terms must be a multiple of 101. The term $20 \times B$ is a multiple of 101 and B is a digit. This happens only for $B = 0$.
The equation simplifies to $A + C = 11 \times D$. Since A and C are digits and D is not 0, $A + C = 11$ and $D = 1$. Notice that pairs of A and C can be $2 + 9 = 11, 3 + 8 = 11, 4 + 7 = 11$, or $5 + 6 = 11$.

20. (A) 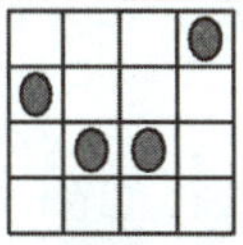

Compare the initial positions of the ladybugs with the positions after the first whistle. There is only one cell that has a ladybug at both times, so that ladybug is the sleeping one.

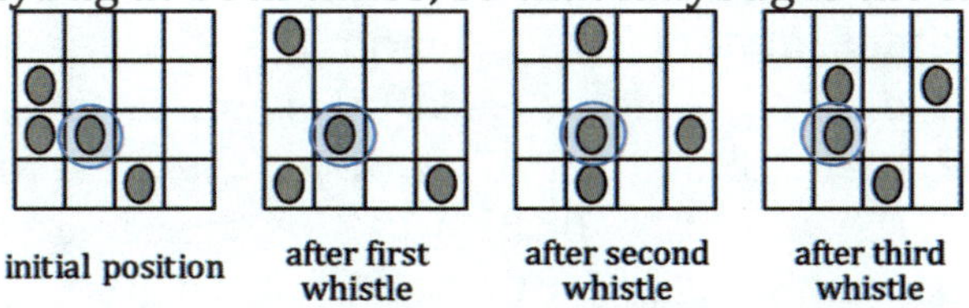

That ladybug needs to stay in the cell even after the fourth whistle, so option (E) cannot happen.
Option (D) is excluded because the ladybug at the very bottom can't move two positions up.
Option (C) is excluded because the ladybug above the sleeping ladybug can't move diagonally to the upper left corner. The ladybugs can only move up, down, right, or left.
Look at the right column after the second whistle, after the third whistle, and also at option (B). The ladybug in the right column in (B) occupies the same position as after the second whistle. However, the ladybugs are not allowed to go back to the cells they came from, so (B) is also excluded.
Only (A) shows a configuration after the fourth whistle that complies with the rules.

21. (C) 2
Neither Masha nor Dasha selected 6 or 7 since the smallest possible sum would be $6 + 1 + 2 = 9$, which is more than 8. Thus, each girl is selecting three numbers from the list 1, 2, 3, 4, 5. If 1 is not selected, then the smallest sum is $2 + 3 + 4 = 9$, which is more than 8. Hence, **both girls must select 1**. After that, Masha is selecting two numbers from the list 2, 3, 4, 5 so that their sum is 7, and Dasha is selecting from the same list two numbers so that their sum is 6. Notice that $2 + 3 = 5$, $2 + 4 = 6$, $2 + 5 = 7$, $3 + 4 = 7$, $3 + 5 = 8$, and $4 + 5 = 9$, so Dasha must select **2** and **4** to get 6. Masha can select either **2** and 5 or 3 and **4** to get 7. In either case, two numbers selected by Masha and Dasha, 1 and either 2 or 4, are the same.

22. (C) C
The 3rd scale is in balance and shows all five balls. The only possible distribution of weights is 80 g + 50 g = 50 g + 50 g + 30 g, so one of B, C, or E weighs 30 g and the other two 50 g each. The 2nd scale shows B and E weighing more than A and C. If either B or E weighs 30 g, then B and E together weigh 50g + 30 g = 80 g, which can't weigh more than any other two balls. Thus, C weighs 30 g, and B and E together weigh 50 g + 50 g =100 g. If A weighs 80 g, then A and C would weigh 80 g + 30 g = 110 g, which is more than 100 g. Therefore, D weighs 80 g, C weighs 30 g, and the other three balls weigh 50 g each. The 1st scale shows that 30 g + 80 g = 110 g weighs more than 50 g + 50 g = 100 g.

23. (D) AAABCB
For any valid form the three As must be together and the two Bs must be together. Otherwise, we can switch two different neighboring digits to get a bigger number. For example, AAABCB becomes AAACBB if B is the smallest or AAABBC if C is the smallest of the three given digits. Suppose that the three different digits are 1, 2, and 3.
If $A = 3$, $B = 2$ and $C = 1$, then 333221 (or AAABBC) is the greatest possible 6-digit number.
If $A = 3$, $B = 1$ and $C = 2$, then 333211 (or AAACBB) is the greatest possible 6-digit number.
If $A = 2$, $B = 3$ and $C = 1$, then 332221 (or BBAAAC) is the greatest possible 6-digit number.
If $A = 2$, $B = 1$ and $C = 3$, then 322211 (or CAAABB) is the greatest possible 6-digit number.
If $A = 1$, $B = 3$ and $C = 2$, then 332111 (or BBCAAA) is the greatest possible 6-digit number.
If $A = 1$, $B = 2$ and $C = 3$, then 322111 (or CBBAAA) is the greatest possible 6-digit number.
Hence, AAABCB is not a valid option.

24. (C) 45
The difference between the granny's age and Kate's age (which is how old her granny was when Kate was born) is the same as the difference between the sum of the granny's and the mother's ages and the sum of Kate's and the mother's ages, because the mother's age is included in both sums. This difference is $81 - 36 = 45$, so the granny was 45 years old when Kate was born.

25. (B) 3
$2 + 3 + 4 + 5 + 6 + 7 + 8 + 9 + 10 = 54$, so a common sum of any number of groups must be a divisor 54. The divisors of 54 are 1, 2, 3, 6, 9, 18, 27, and 54. The common sum can't be smaller than 10 (10 is a number on the list) and the number of groups is the largest when the common sum is as small as possible. The best option is 18 with three groups. This actually can be done since $3 + 4 + 5 + 6 = 18$, $2 + 7 + 9 = 18$, and $8 + 10 = 18$.

26. (D) 200 cm

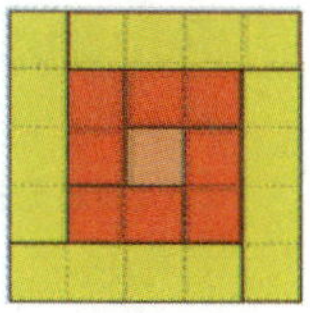

All 9 pieces have the same width of 8 cm since the board was 8 cm wide, so the side length of the square at the center is 8 cm. Around this square there are 4 rectangles each with the length of 2×8 cm and 4 outer rectangles each with the length of 4×8 cm. We put these 9 pieces one after another in one row with the combined length of $8 \text{ cm} + 4 \times (2 \times 8 \text{ cm}) + 4 \times (4 \times 8 \text{ cm}) =$
$= (1 + 8 + 16) \times 8 \text{ cm} = 25 \times 8 \text{ cm} = 200 \text{ cm}$. Hence, the board was 200 cm long.

27. (B) 21

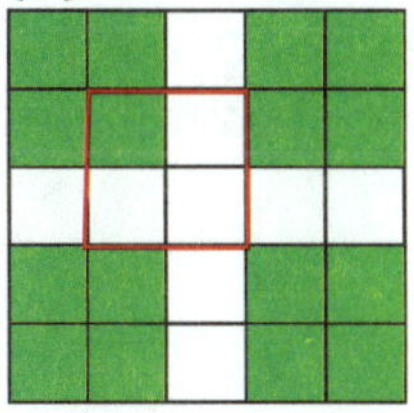

Each of 2×2 green squares must contain at least one 0 and the four squares do not share cells, so the 5×5 table contains at least four 0s. To have the greatest sum, we must have as many 1s and as a few 0s as possible. If there are no 0s outside the green area, then the 2×2 square with red perimeter must share its 0 with the green square it overlaps. The same argument works for other green squares suggesting the arrangement of 0s and 1s shown to the right. Each 2×2 square of this table contains exactly 3 equal numbers and the table must contain at least four 0s, so we can't have more than 21 ones. Therefore, 21 is the largest possible sum of all the numbers in the table.

1	1	1	1	1
1	0	1	0	1
1	1	1	1	1
1	0	1	0	1
1	1	1	1	1

28. (C) 9

A person sitting between two liars is telling the truth since both neighbors of that person are liars. A person seating between two people telling the truth is a liar. Also, a person sitting between a liar and a person telling the truth is a liar since one of the neighbors is not a liar. You can't have two neighbors both telling the truth since they would contradict each other about having both their neighbors as liars. You can't have a group of three liars since the person in the middle would be telling the truth. However, you can have two liars as neighbors.

To have as many liars as possible create, according to the above rules, as many pairs of lying neighbors as possible. You can have 4 such pairs and an additional single liar seating between people telling the truth, so the maximum number of liars at the table is 9 as shown to the right.

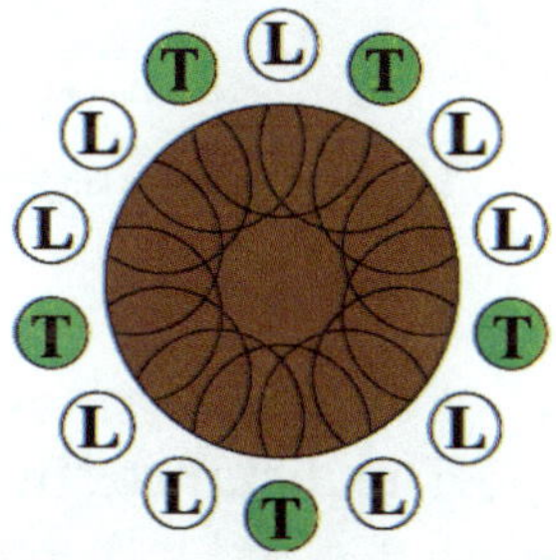

29. (C) 3

If we arrange 8 domino tiles into a 4×4 square so that the number of dots in each row and column is the same, that number multiplied by 4 is the number of all the domino dots (we count them either row by row or column by column).

The domino tiles show 4×1 dot, 4×2 dots, 2×3 dots, 2×4 dots, 1×5 dots, and 1×6 dots. The sum is $4 \times 1 + 4 \times 2 + 2 \times 3 + 2 \times 4 + 1 \times 5 + 1 \times 6 = 37$ dots but one number of dots is not shown, so 37 + the missing number must be a multiple of 4. Among the options $37 + 0 = 37$, $37 + 1 = 38$, $37 + 2 = 39$, $37 + 3 = 40$, $37 + 4 = 41$, $37 + 5 = 42$, and $37 + 6 = 43$, only $37 + 3 = 40$ is a multiple of 4, so 3 dots are on the covered part. The number of dots in each row and in each column must be $40 \div 4 = 10$. The required arrangement of domino tiles actually exists as shown to the right.

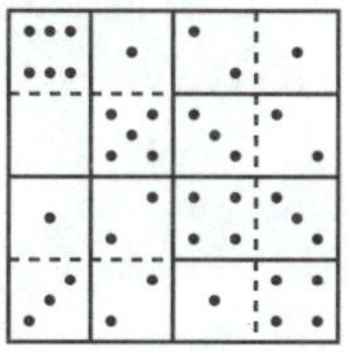

30. (E) 18

For a particular distribution of the seven numbers, we add the three sums along the three straight lines, which is the same as adding all the numbers from 3 to 9 plus twice the number in the center circle (the center circle is included along all three lines, so it is added two more times). The three sums are equal, so when we add them the result must be a multiple of 3. Notice that $3 + 4 + 5 + 6 + 7 + 8 + 9 = 42$, so ($42 + 2 \times$ the number in the center circle) must be a multiple of 3. Among the numbers $42 + 2 \times 3 = 48$, $42 + 2 \times 4 = 50$, $42 + 2 \times 5 = 52$, $42 + 2 \times 6 = 54$, $42 + 2 \times 7 = 56$, $42 + 2 \times 8 = 58$, and $42 + 2 \times 9 = 60$, only $42 + 2 \times 3$, $42 + 2 \times 6$ and $42 + 2 \times 9$ are multiples of 3. So, no numbers other than 3, 6, or 9 can be placed in the circle with the question mark. $3 + 6 + 9 = 18$ is the answer if there are actual arrangements with each of the three numbers at the center. They exist and are shown below.

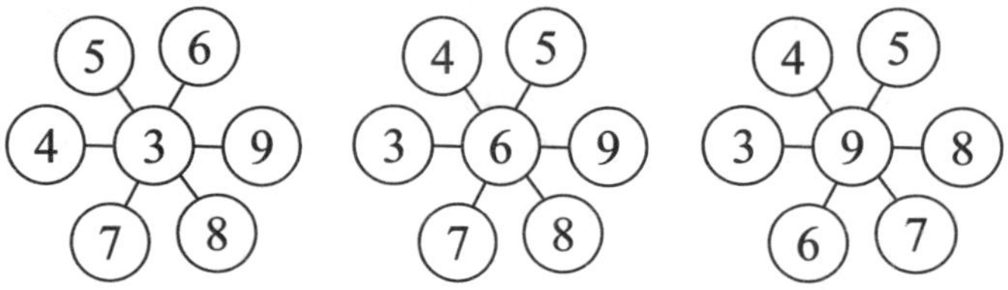

Solutions for Year 2020

1. (E) 

We need a piece with one quadrant of the circle and two narrow kite-shaped spikes at opposite corners. Only (E) has this property and it also matches the fourth corner of the square with the question mark.

2. (E) 

For each signpost we add the two numbers to get the distance between Atown and Betown. The distances in kilometers in the order from (A) to (E) are: $2 + 9 = 11$, $3 + 8 = 11$, $5 + 6 = 11$, $8 + 3 = 11$, and $9 + 4 = 13$. All signposts except the signpost (E) have the distance of 11 km between the two towns, so the signpost (E) is incorrect.

3. (D) 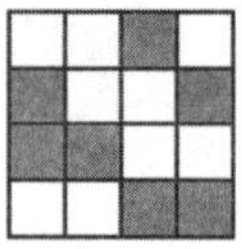

The original square has the following colors (listed row by row): GGWG, WGGW, WWGG, GGWW. G stands for gray and W stands for white. After switching the colors, the rows will be WWGW, GWWG, GGWW, WWGG. This describes the square in (D).

4. (B) 2

A pair of eggs is needed to bake 6 muffins. A box of eggs contains 3 pairs of eggs, which is enough to bake $3 \times 6 = 18$ muffins. Mikas wants to bake 24 muffins, so he needs another pair of eggs. For this reason, he has to buy another box of eggs. Hence, Mikas needs to buy 2 boxes of eggs to make 24 muffins.

5. (E) 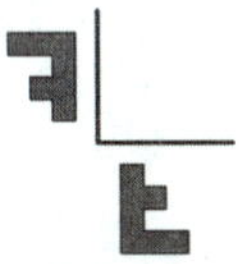

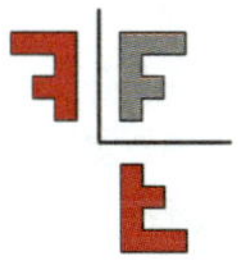

The reflection of F with respect to the vertical line is  and the reflection of F with respect to the horizontal line is , so (E) is the answer.

6. (C) 13

$10 = 5 + 5$, $12 = 5 + 7$, $14 = 7 + 7$, and $15 = 5 + 5 + 5$, so Kim can create the lengths 10, 12, 14, and 15 by joining various chains of the lengths 5 and 7 one after another. She cannot create the length 13.

7. (A) 3

Maria starts with 10 pieces of paper (all of them are sheets). If she cuts 1 sheet into pieces, she has $1 \times 5 + (10 - 1) = 14$ pieces of paper. If she cuts 2 sheets into pieces, she has $2 \times 5 + (10 - 2) = 18$ pieces of paper. If she cuts 3 sheets into pieces, she has $3 \times 5 + (10 - 3) = 22$ pieces of paper. Thus, Maria cut 3 sheets.

8. (B) 3

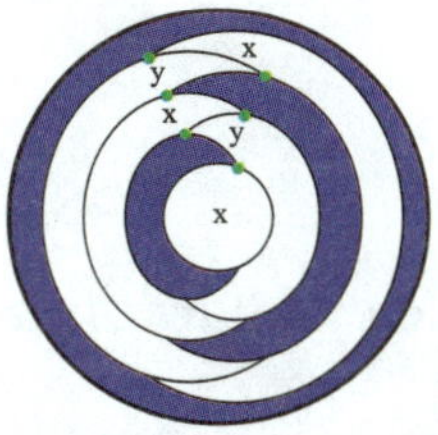

There are 8 regions and 6 points where 3 regions meet. These points are marked in green. The 3 regions at each point must be 3 different colors. The outer region is blue, and x and y stand for the other two colors. In the completed pattern there are 3 blue regions. If we choose red for x, then there are 3 red regions and 2 yellow regions. If we choose yellow for x, then there are 2 red regions and 3 yellow regions. However, there are always 3 blue regions.

9. (C) 5

There are $1 + 4 + 6 + 9 = 20$ apples. To distribute them evenly among 4 baskets, we have to have 5 apples in each basket. We have to add 4 apples to the first basket and 1 apple to the second basket, so we have to move at least $4 + 1 = 5$ apples. You can move 4 apples from the fourth basket to the first one and 1 apple from the third basket to the second one to get the 20 apples evenly distributed.

10. (E) at E

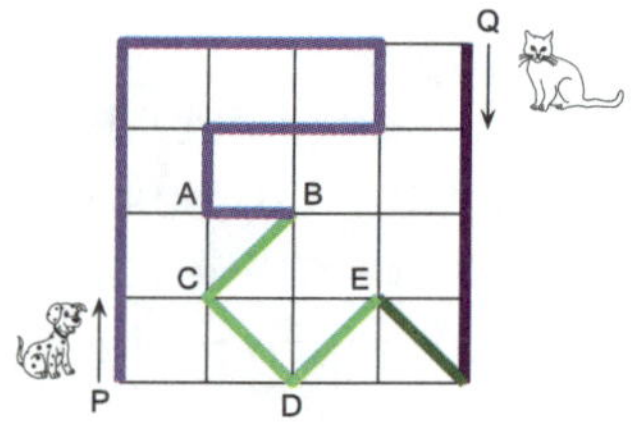

When the dog moves from P to B, it covers 12 vertical and horizontal segments of the same length. The cat moves down through 4 such segments and 12 : 4 is the same as 3 : 1. The dog walks 3 times as fast as the cat, so the dog is at B when the cat is at the bottom of the right edge of the square. At this moment the dog and the cat are separated by 4 equal diagonal segments of the path. To keep the proportion 3 : 1, the dog walks 3 diagonal segments and the cat walks 1 such segment and they meet at the point E.

11. (A) 3

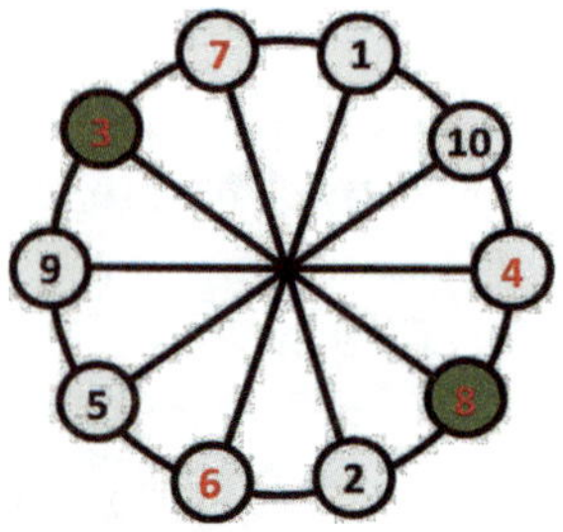

The circles diametrically opposite the circles with 1 and 10 are the circle with 5 and the one counterclockwise to it. 6 must be placed in that circle as $10 + 1 - 5 = 6$.
By the same rule $6 + 2 - 1 = \mathbf{7}$ must be placed in the circle to the left of the circle with 1. Similarly, to the right of the circle with 10 we have to place $5 + 9 - 10 = \mathbf{4}$.
So far, numbers 1, 2, 4, 5, 6, 7, 9, and 10 have been placed in the corresponding circles, so in the circle with the question mark we can place either 8 or 3. If 8 is placed there, then in the opposite circle we have to place $7 + 8 - 2 = 13$, which is not a valid number. Thus, 3 must be placed in the circle with the question mark and $7 + 3 - 2 = 8$ must be placed in the opposite circle.

12. (E) 5 hours and 42 minutes

To see **20:20** when hanging upside down, the digital clock must be showing **02:02** (rotation by 180°). There are 3 hours and 40 minutes from 20:20 to midnight and 2 hours and 2 minutes from midnight to 02:02. 3 h 40 min + 2 h 2 min = 5 h 42 min, so Elise the bat was away from her cave for 5 hours and 42 minutes.

13. (A) I am telling the truth.

Both a liar and a truth-teller can say, "I am telling the truth," since in the case of a liar it is a lie and in the case of a truth-teller it is the truth.
Neither of them can say, "You are telling the truth," since a liar could not tell the truth about a truth-teller (or anything) and a truth-teller could not tell a lie about a liar (or anything else).
A truth-teller can't say, "We both are telling the truth," since it is not true as a liar is not telling the truth.
Neither of them can say, "I always lie," since a truth-teller is always telling the truth and a liar can't make a true statement.
A liar can't say, "One and only one of us is telling the truth," since it would be a true statement.
Therefore, (A) is the only statement that can be said by both a liar and a truth-teller.

14. (B) 

Cube (A) has 9 visible dark gray cubes and Mary has only 8 dark gray cubes, so she didn't build cube (A). Cube (C) has 11 visible white cubes and Mary has only 10 white cubes, so she didn't build cube (C). Cube (D) has 9 visible dark gray cubes and Mary has only 8 dark gray cubes, so she didn't build cube (D). Cube (E) has 10 visible light gray cubes and Mary has only 9 light gray cubes, so she didn't build cube (E). The only cube left is (B). It has 19 visible cubes, 10 of them are white (Mary has 10 white cubes) and 9 of them are light gray (Mary has 9 light gray cubes). 8 cubes of (B) are not visible and Mary has exactly 8 dark gray cubes, so only cube (B) can match Mary's set of cubes. She built one of the 5 big cubes, so she built the big cube (B).

15. (C) 

Look at all 5 paths. Each path has 5 vertical/horizontal segments of the same length and 2 medium circular segments. The number of large circular segments and the number of small circular segments differ. From (A) to (E) these numbers are: 2L + 2S, 2L + S, L + 2S, 2L + 2S, and 2L + S, where L and S stand for the lengths of large and small circular segments, respectively. L > S, so L + 2S is the smallest of these five lengths. Therefore, (C) is the shortest path.

16. (C) 7
Right now the father has 36 votes and the children have $13 + 6 + 4 = 23$ votes. Each year father's age increases by 1 and children's combined age increases by 3. 6 years from now the father will have $36 + 6 = 42$ votes and the children will have $(13 + 6 + 4) + 3 \times 6 = 23 + 18 = 41$ votes. 7 years from now the father will have $36 + 7 = 43$ votes and the children will have $(13 + 6 + 4) + 3 \times 7 = 23 + 21 = 44$ votes, so after 7 years the children will have majority of the votes.

17. (E) 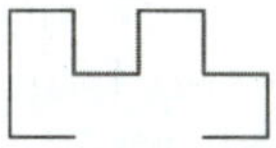

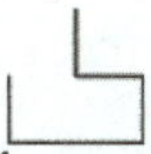 is the shape of two identical pieces of wire. The shape consists of 5 linear segments. 4 segments have the same length and the fifth segment has double that length. Therefore, two pieces of wire put together must have at least 2 segments of double length. (E) has only one segment of double length, so (E) cannot be made by putting together these two pieces.
Here are shapes (A) through (D):

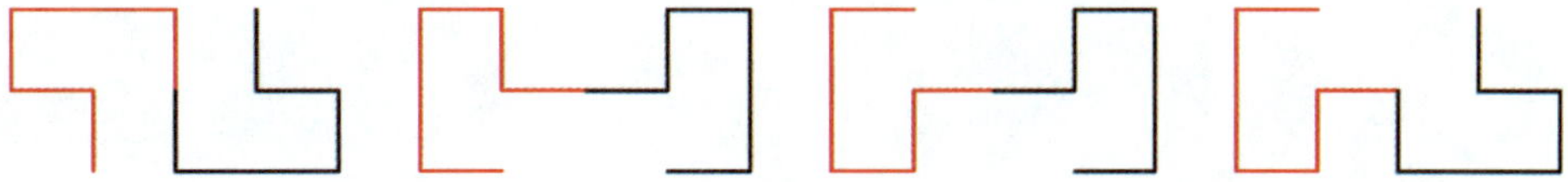

18. (D) 

One view of the cube shows the ladybug's head towards the edge adjacent to the face with the dog.

The other view shows the head of the ladybug towards the edge adjacent to the bottom face, so the dog is on the bottom face.

Hence, the sticker with the dog is on the face opposite the face with the mouse.

19. (B) 3

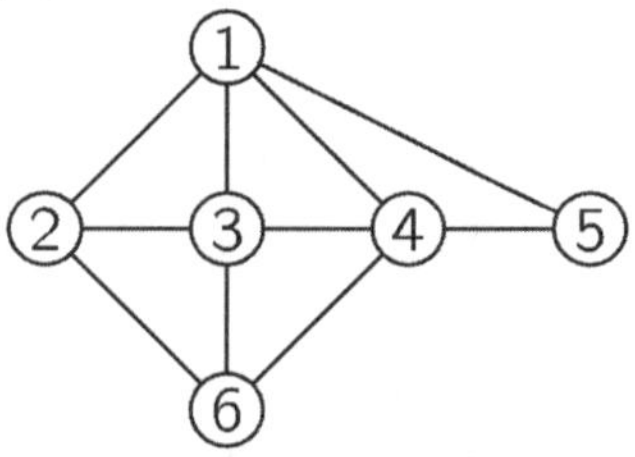

① has 4 friends, ② has 3 friends, ③ has 4 friends, ④ has 4 friends, ⑤ has 2 friends, and ⑥ has 3 friends, so there is only one girl with 2 friends and Beatrice is that girl. She is represented by ⑤. Her friends are Chloe and Diana, so they are associated (in some order) with ① and ④. Except ① and ④, only ③ has 4 friends, so ③ represents Fiona.

20. (A) 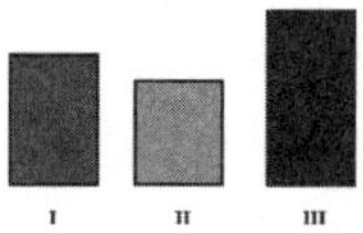

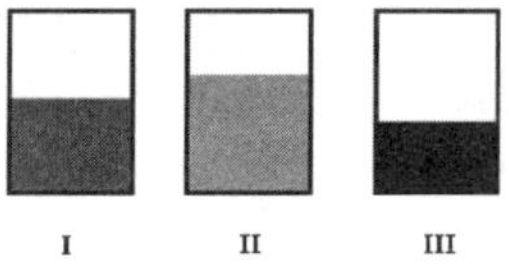

The front faces of three rectangular vessels are shown to the right, so all the vessels have the same widths and the same heights. The level of liquid in each vessel is indicated by its shaded area and the amount of liquid in each vessel equals its shaded area × its length (the third dimension of the vessel, not visible but going away from the viewer in the original picture).
All the vessels contain the same amount of liquid. That puts the lengths of the vessels in reverse order of their shaded areas, so the order of the lengths should be (III) > (I) > (II).
The different lengths (and identical widths) are visible when looking at the vessels from above. The length of each vessel is represented by the vertical dimension in the view of the answer choices. Only (A) shows these vertical segments in the order (III) > (I) > (II), so (A) represents the three vessels when viewed from above.

21. (B) 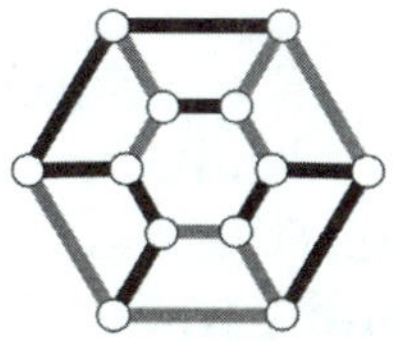

When we look at the frustum (truncated pyramid), we see 6 side faces as identical isosceles trapezoids. To the left of the front face there is one special face, the only face with the parallel edges black and the other two edges gray. This excludes options (C) – (E) as valid answers. (A) is excluded since it has a trapezoid with 4 black edges and there is no such side face in the frustum. In diagram (B) the special face is shown at the top. We can match all the trapezoids to the side faces of the frustum when moving counterclockwise from the special face towards the front face.

22. (E) 19 cm

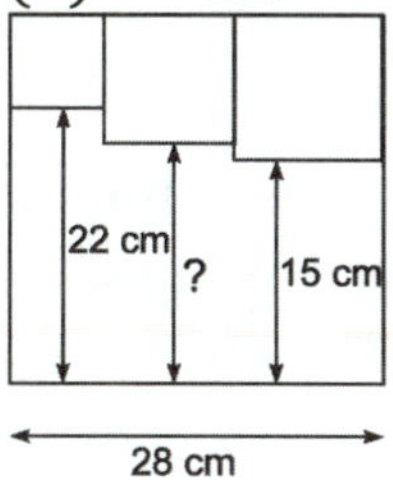

There are three smaller squares inside the big square of the size 28×28. The side of the square at the left corner is 28 cm – 22 cm = 6 cm long. The side of the square at the right corner is 28 cm – 15 cm = 13 cm long. Thus, the side of the square in the middle is 28 cm – (6 cm + 13 cm) = 9 cm long. Hence, the length of the line marked with the question mark is 28 cm – 9 cm, which is 19 cm. Notice that $28 - [28 - (6 + 13)] = 6 + 13 = 19$.

23. (B) 2

Each token is white on one side and black on the other side. If you flip any three tokens, then at least 1 black token and at least 2 white tokens are still there, so we need at least 2 turns to have all tokens of the same color.
It can be done in 2 turns. In the first turn flip any 2 black tokens and 1 white token. After that $(2 + 1) = 3$ tokens are black and $(4 + 2) = 6$ tokens are white. In the next turn flip the 3 black tokens, so after these 2 turns all 9 tokens are white. Notice that to turn all tokens black we need at least 3 turns.

24. (C) △○○○

The two balanced scales are  and .

"Add" them to see the new balanced scale 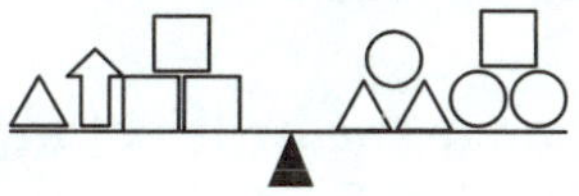.

Remove one triangle and one square from each side to see the next balanced scale 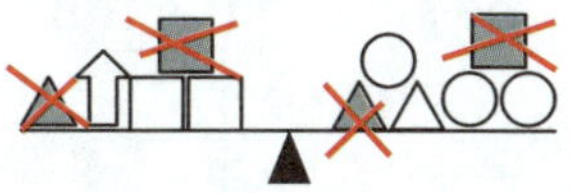.

The left-hand side of the above scale matches the left-hand side of the third scale (with the question mark). (C) matches the right-hand side of the above scale, so (C) definitely balances the third scale

25. (D) lemon with a wafer

There are 4 scoops of vanilla and only 4 different kinds of toppings. To avoid any duplication, we have to use 1 umbrella, 1 cherry, 1 wafer, and 1 chocolate chip on the 4 vanilla scoops. After this we are left with 3 scoops of chocolate and 3 different kinds of toppings (the chocolate chip is gone). To avoid any duplication, we have to use 1 umbrella, 1 cherry, and 1 wafer on the 3 scoops of chocolate. After these two steps we are left with 2 scoops of lemon and 2 different kinds of toppings (the chocolate chip and wafers are gone). To avoid any duplication, we have to use 1 umbrella and 1 cherry on the 2 lemon scoops. After these three steps we are left just with 1 scoop of mango and 1 topping of umbrella (the chocolate chip, wafers, and cherries are gone), so the last umbrella goes on top of the scoop of mango. In summary, we have the following ten matches: vanilla – umbrella (C), vanilla – cherry, vanilla – wafer, and vanilla – chocolate chip (E); chocolate – umbrella, chocolate – cherry (A), and chocolate – wafer; lemon – umbrella and lemon – cherry; and mango – umbrella (B).
Among the 5 options listed as possible answers only (D) lemon with a wafer is not shown.

26. (D) 8

Any positive integer that ends with 0 is preceded by an integer that ends with 9. Any number ending with 9 can't be a *nice* number. To get the longest sequence of consecutive *nice* numbers, the middle digit must be as large as possible (which is 9) and the first digit as small as possible (which is 1). The longest sequence of consecutive *nice* numbers is 190, 191, 192, 193, 194, 195, 196, 197. This sequence has 8 terms.

27. (B) 3

$\frac{1}{2}+\frac{1}{3}+\frac{1}{6}=1$, so $\frac{m}{2}+\frac{m}{3}+\frac{m}{6}=m$, where m is the number of games played by Magnus until $\frac{m}{2}+\frac{m}{3}+2=m$. It happens when $\frac{m}{6}=2$ or $m=12$. It makes sense since $12<15$, $\frac{12}{2}=6$, $\frac{12}{3}=4$, and $6+4+2=12$ (6 wins, 4 losses, and 2 draws). Magnus still has $15-12=3$ games left to play.

28. (A) $a=6, b=4, c=8$

1		7
2		8
3		9

After the first folding, 1 is on the top of the first column and 9 is at the bottom of the third column, so the square to the left shows the numbers in the proper order after the first folding. We also see that the top row can be marked as

1	a	7

.

The second folding doesn't change the order of numbers in the first column and it doesn't change the order of numbers in the third column. It puts the cell with a directly on top of the cell with 7, so $a=6$. In the square the numbers under 6 in the middle column are 5 and 4.

Here is the complete solution

1	6	7
2	5	8
3	4	9

. Compare it to

1	a	
		b
	c	

.

Thus, $a=6, b=4$, and $c=8$.

29. (E) 36

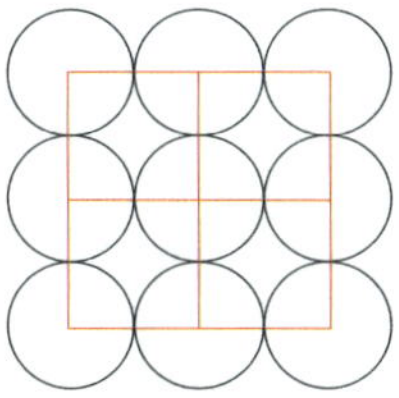

The contact points of the 3×3 balls of the base are in the plane passing through their centers (shown in the diagram to the left).
There are 3×2 contact points along the horizontal lines and 3×2 contact points along the vertical lines, so there are 12 contact points among the 3×3 balls of the base.

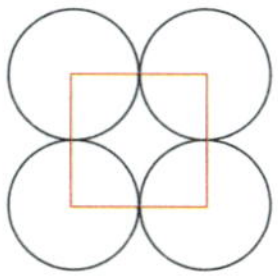

The diagram to the right shows that there are 4 contact points among the 2×2 balls of the middle layer.
Each of the 2×2 balls is touching 4 balls underneath and all these contact points are different, so there are $4 \times 4 = 16$ such contact points.
The top ball is touching each of the 2×2 balls underneath, so there are 4 more contact points.
Altogether, there are $12 + 4 + 16 + 4 = 36$ contact points.

30. (B) east

Do not attempt to draw the whole path in one go, by trial and error. The idea is to gradually discover parts of the path. For example, there are some islands that are connected to exactly two others, one on each side. There is only one path through these islands. Ignoring the islands marked with "start" and "finish," we marked these islands and the paths that necessarily go through them in green on the map below. Now it is easy to complete the path. It has only one solution, as shown in red on the right. So, from the island in the middle the postman must move east.

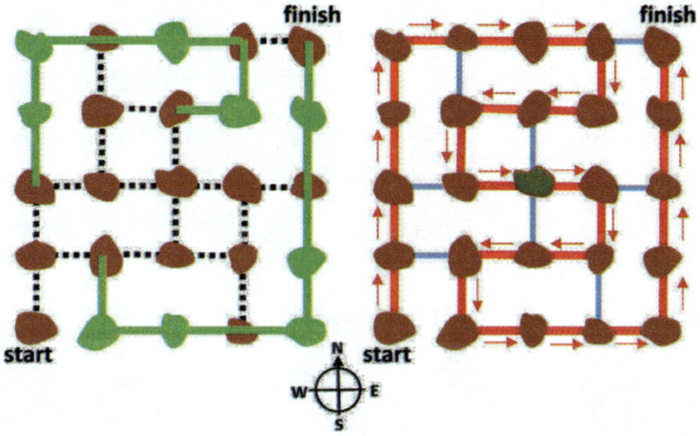

Part III: Answer Keys

	1998	2000	2002	2004	2006	2008
1	B	C	B	C	D	C
2	A	A	C	C	E	C
3	B	D	D	E	D	B
4	C	E	E	E	B	D
5	A	D	C	E	E	E
6	E	E	D	D	A	C
7	C	D	D	D	D	B
8	D	B	C	B	A	E
9	D	E	E	C	C	A
10	C	E	B	D	D	E
11	B	B	C	D	D	E
12	B	E	A	E	D	B
13	D	D	C	D	E	D
14	C	B	E	B	A	E
15	D	C	E	B	B	D
16	C	E	C	B	A	D
17	D	D	B	E	B	C
18	A	E	A	C	C	C
19	C	D	A	C	B	B
20	D	D	C	D	B	C
21	D	D	C	C	E	E
22	E	D	B	B	B	C
23	C	B	D	B	D	D
24	C	B	A	A	C	C
25	B	A	A	C	E	D
26	E	A	A	A	D	C
27	B	D	A	D	A	D
28	C	E	B	D	B	D
29	D	D	C	A	E	D
30	A	A	A	C	C	C

	2010	2012	2014	2016	2018	2020
1	B	C	C	C	A	E
2	D	C	D	E	C	E
3	C	A	D	A	B	D
4	C	C	A	C	B	B
5	C	E	A	E	D	E
6	B	C	D	B	E	C
7	B	D	B	D	D	A
8	B	A	B	C	C	B
9	E	B	B	C	C	C
10	C	B	C	B	A	E
11	E	D	E	A	E	A
12	E	B	B	C	B	E
13	D	D	D	C	E	A
14	D	C	B	D	A	B
15	D	D	B	B	A	C
16	C	D	D	E	C	C
17	A	C	A	A	C	E
18	C	D	D	D	E	D
19	C	A	E	A	A	B
20	E	C	A	C	A	A
21	D	B	D	D	C	B
22	C	D	A	B	C	E
23	B	D	E	D	D	B
24	B	B	C	B	C	C
25	E	D	E	B	B	D
26	B	B	D	D	D	D
27	E	D	B	D	B	B
28	B	D	C	E	C	A
29	D	C	C	C	C	E
30	C	B	E	E	E	B